MW00809458

Remanufacturing in the Circular Economy

Scrivener Publishing
100 Cummings Center, Suite 541J
Beverly, MA 01915-6106

Publishers at Scrivener
Martin Scrivener (martin@scrivenerpublishing.com)
Phillip Carmical (pcarmical@scrivenerpublishing.com)

Remanufacturing in the Circular Economy

Operations, Engineering and Logistics

Edited by
Nabil Nasr
Golisano Institute for Sustainability at Rochester Institute of Technology, New York, USA

This edition first published 2019 by John Wiley & Sons, Inc., 111 River Street, Hoboken, NJ 07030, USA and Scrivener Publishing LLC, 100 Cummings Center, Suite 541J, Beverly, MA 01915, USA

For more information about Scrivener publications please visit www.scrivenerpublishing.com.

Wiley Global Headquarters
111 River Street, Hoboken, NJ 07030, USA

For details of our global editorial offices, customer services, and more information about Wiley products visit us at www.wiley.com.

Library of Congress Cataloging-in-Publication Data

Names: Nasr, Nabil (Engineering professor), editor.
Title: Fundamentals of remanufacturing : operations, engineering and logistics / edited by Nabil Nasr, Golisano Institute for Sustainability at Rochester Institute of Technology, New York, USA.
Description: First Edition. | Hoboken, New Jersey : John Wiley & Sons, Inc. ; Salem, Massachusetts : Scrivener Publishing LLC, 2020. | Includes bibliographical references and index.
Identifiers: LCCN 2019034788 (print) | LCCN 2019034789 (ebook) | ISBN 9781118414101 (hardback) | ISBN 9781119664345 (adobe pdf) | ISBN 9781119664376 (epub)
Subjects: LCSH: Remanufacturing.
Classification: LCC TS183.8 .F86 2020 (print) | LCC TS183.8 (ebook) | DDC 670--dc23
LC record available at https://lccn.loc.gov/2019034788
LC ebook record available at https://lccn.loc.gov/2019034789

Cover image: Pixabay.com
Cover design by Russell Richardson

Set in size of 13pt and Minion Pro by Manila Typesetting Company, Makati, Philippines

10 9 8 7 6 5 4 3 2 1

Contents

Preface

Over the past thirty years or so, I've discovered firsthand that it can be difficult to explain the fundamentals of remanufacturing in simple, succinct ways. Remanufacturing is a unique field because it draws upon a broad array of skillsets, technologies, processes, and business models. It is highly technical as an industrial system while also deeply conceptual when approached through a sustainability lens. All this makes describing it to the many business leaders, engineers, and policymakers that I have worked with across the globe no easy task, as they all have different perspectives and levels of expertise across varied disciplines.

As an engineer, I am drawn to a problem in need of a solution. This is why I am so pleased to introduce a book that will present remanufacturing fundamentals to a wide audience in an accessible, useful way. The contributors to this book present a diverse set of expertise, bringing significant knowledge to the topic areas they cover. Of course, any project that attempts to capture the whole of a complex, fast-changing area is going to run into barriers and this book is no exception.

Introduction to Remanufacturing

Remanufacturing is an industrial process that restores used, worn and retired products or modules to a like-new condition. The restoration is typically a highly engineered process done in an industrial setting through which products are systematically disassembled, cleaned, and inspected for wear and degradation. Damaged or degraded components are either restored to their original specifications or replaced, feature upgrades can be incorporated, and the product is reassembled. Finally, reliability and quality testing are performed to ensure that performance meets original product specifications.

Remanufacturing has inherent benefits over recycling. Recycling reduces products into raw material which can then be used again. This in fact brings significant environmental and possibly economical value. However, in contrast, remanufacturing retains the geometrical shape and function of the products and components, and is therefore able to capture both the materials and the value added (the labor, energy, and manufacturing processes) which had become embodied in the original product.

Recapturing the value-added component of a product is often both environmentally and economically beneficial. In fact, remanufacturing, while very different than recycling, is oftentimes referred to as the "ultimate form of recycling" because it is able to preserve the embodied energy and value add contained in a product. Typically, at the end of its service life, a product would be destined for landfill, incineration or recycling. But by implementing a remanufacturing strategy, disposal costs can be avoided, the value still embodied in a product can be recouped, and resources can be used more efficiently. In addition, a fact that is often not mentioned, remanufacturing is a gateway for recycling, as when end of service products, called cores, are returned to a remanufacturing facility, products are disassembled and components often properly separated and identified. The components that are not suitable for reuse in the remanufacturing process are then properly sent for recycling with significantly better separation of like materials, thus significantly improving the recycling recovery and output.

Growth of an Industry

It is widely accepted that the United States has the most diverse and developed remanufacturing industry sectors in the world.

Much of this development stems from the industrial revolution and the subsequent invention of mass production. A good example can be seen by looking at the history of Ford Motor Company. World War II brought significant material availability challenges which resulted in significant growth for the remanufacturing industry due to the significantly lower virgin material needed and high reuse rate of used materials. Once Ford had produced and sold millions of Model T cars in the early 1900s, it made practical business sense to create separate facilities to remanufacture old motors and a distribution system to support their customers with economical like-new engines and service parts.

This pragmatic approach to integrating remanufacturing into global product support strategies continues to this day. Thousands of product categories including aerospace, automotive, military systems, office furniture and equipment, transportation, construction and electrical equipment; medical devices; machine tools, compressors, other heavy machinery and others account for the global remanufacturing business activity. Major corporations such as Caterpillar, John Deere, Xerox and General Electric all generate significant business through remanufacturing programs related to their products. Conservative estimates show more than $60 billion of remanufactured goods are sold each year in the United States alone and more than 500,000 people are employed by the remanufacturing industry.

Remanufacturing's Environmental Benefits

Many manufacturers have begun looking for new ways to increase efficiency and reduce costs while developing manufacturing processes that reduce or eliminate negative environmental impacts. This interest in sustainable production inevitably leads these organizations to explore the opportunities and benefits of reuse, remanufacture, and recycling of manufactured goods and products.

Remanufacturing is an important factor in sustainable production because it recovers and preserves much of the expensive value-added component of a manufactured product. Through remanufacturing, nonrenewable resources are kept in circulation for multiple product lifecycles, thus conserving up to 80% or more of the original raw materials, labor and energy embedded in the product. According to an Argonne National Laboratory study, remanufactured products conserve the equivalent of 400 trillion BTUs of energy annually, enough to power 6 million passenger cars each year. This also avoids the generation of 28 million tons per year of the greenhouse gas CO2. These key environmental benefits are readily seen in data from remanufacturing market leaders such as Caterpillar.

Caterpillar is one of the world's largest remanufacturers, processing more than two million units annually and recycling more than 100 million pounds of remanufactured products each year. The company offers remanufacturing services for a variety of products and components to serve Caterpillar and external clients. Their Components business includes undercarriage, ground engaging tools, hose and connectors, hardened bar stock, tubes, specialized products, common components, fluids and filters. Caterpillar points to an engine cylinder head to illustrate the sustainability of remanufacturing. Compared with manufacturing a new cylinder head, a remanufactured head requires 61% less greenhouse gas, 93% less water, 86% less energy, up to 99% less material use, and it contributes up to 99% less space in a landfill.

Remanufacturing Industry Sectors

Remanufacturing activity encompasses thousands of product categories, from cell phones and laptops to jetliner engine parts to armored ground combat vehicles. However, the majority of remanufacturing industry activity is concentrated in 13 large U.S. sectors, which are described in detail in this book.

After decades of relative obscurity, the remanufacturing industry has emerged larger, more diverse, and a bigger contributor to national and global economies than most of us realize or appreciate. And as all industries transition to greater sustainability in the decades ahead, remanufacturing will be an increasingly more powerful driver for change. So, I invite you to become better acquainted with the scope of remanufacturing in America and around the globe.

Acknowledgments

Nothing convinces people so much as living proof. Caterpillar, Inc., is among the foremost remanufacturers in the world. They are committed to building a global remanufacturing infrastructure to enable progress across industries. This book would not have been possible without the support they made possible through the Caterpillar Professorship and the Caterpillar Fund at the Golisano Institute for Sustainability at Rochester Institute of Technology (RIT), where I serve as associate provost and director.

Nabil Nasr
Golisano Institute for Sustainability at
Rochester Institute of Technology, New York, USA
July 2019

1

Value-Retention Processes within the Circular Economy

Jennifer Russell[1] and Nabil Nasr[2*]

[1]Virginia Tech, Blacksburg, Virginia, USA
[2]Golisano Institute for Sustainability at Rochester Institute of Technology, New York, USA

Abstract

The circular economy offers a framework for transforming wasteful and inefficient linear systems into cascading systems that retain the inherent value of products, reduce negative externalities, and improve resource-efficiency. The cycling of technical nutrients within a circular economy can be achieved through product value-retention processes (VRPs) that include direct reuse, repair, refurbishment, and remanufacturing. Product case studies reveal that VRPs offer differing degrees of process and resource-use intensity, and as such, each contributes different economic and environmental benefits and circularity. Value-retention and impact metrics, measured relative to new product options, include new material use (kg/unit), energy use (MJ), emissions (kg CO_2-eq.), production waste (kg/unit), cost advantage (% $USD/unit), and employment opportunity (Full-time Laborer/unit or FTE/unit). When compared to a traditional new product, all VRPs create significant resource efficiency and circularity opportunities. When compared to other VRPs, Partial Service-Life VRPs (direct reuse and repair) require significantly fewer resources,

Corresponding author: nzneie@rit.edu

Nabil Nasr (ed.) Remanufacturing in the Circular Economy
(1–30) © 2020 Scrivener Publishing LLC

and thus result in relatively lower environmental and economic costs than Full Service Life VRPs (refurbishment and remanufacturing); However, more intensive Full Service Life VRPs ensure relatively greater utility, service-life, and value for the customer. Because of these differences, VRPs may be adopted strategically to pursue a range of business and policy objectives.

Keywords: Circular economy, value-retention processes (VRPs), resource efficiency, market transformation, remanufacturing, refurbishment, repair, direct reuse

1.1 Introduction

The full potential value of the circular economy goes beyond the recycling of materials in their raw form; in the circular economy, value is ultimately embedded in our ability to retain the embodied and inherent value of product material, structural form, and ultimate function. Capturing, preserving, and re-employing this value not only offsets virgin material requirements, but also reduces required production activities and instills new value altogether by ensuring the completion of, and/or potentially extending a product's expected life. In this respect, value-retaining production processes that include **arranging direct reuse, repair, refurbishment, comprehensive refurbishment, and remanufacturing** (hereafter referred to as value-retention processes or VRPs) are essential for improving industrial system circularity.

Through the deployment and scaling of VRPs worldwide, important environmental and economic objectives of increased system circularity, and the decoupling of economic growth from environmental degradation, can be successfully pursued. There is no single solution that is at once universally applicable, socially equitable, economically efficient, and environmentally healthy. As such, it is critical, to understand the different ways in which these processes may interact within and affect categorically diverse economies.

The International Resource Panel (IRP), a branch of the United Nations Environment Programme (UNEP) investigated each of these VRPs, including their role in the current industrial paradigm, and their potential to impact the future of the circular economy [1]. This assessment helped to shed light on the contribution that VRPs can make to the pursuit of enhanced resource efficiency and the reduction of environmental impacts associated with primary material production and traditional linear manufacturing. Some of the major insights and outcomes of this IRP Report are covered within this chapter.

1.2 Overview and Evaluation of Value-Retention Processes

VRPs are distinctively different from, and far less understood than recycling. VRPs help to ensure the offset of virgin material requirements, the collection and reuse of valuable materials, and the retention of embodied and inherent value, by ensuring the completion of, and/or potentially the extension of a product's expected service life. Expanding the use of VRP practices can offer substantial and verifiable benefits in terms of resource efficiency, circular economy, and protection of the global environment. However, their intensities and adoption globally have been limited due to significant technical, market infrastructure, and policy barriers.

1.2.1 Defining Value-Retention Processes

One of the main challenges facing VRPs around the world, as corroborated via international market access negotiations [2] and the US International Trade Commission (USITC) [3], is the wide range of definitions and interpretations of different VRPs. There are often multiple issues at stake, including common terminology differentiations made within and across sectors, as well as regulations focused on protecting consumer interests in certain countries. For example, while the VRP activity called '*reconditioning*' in the electronics industry (as preferred by the Professional Electrical Apparatus Recyclers League), '*rebuilding*' by the Federal Trade Commission, and '*remanufacturing*' under a definition accepted by the WTO, the intent for each of these terms is the same: "*...the process of returning the electrical product to safe, reliable condition...*" [4]. Alternately, the medical sector typically uses the term '*refurbishment*' for the same VRP that the aerospace sector would use the term '*overhaul*' to describe; In fact, both definitions are clearly describing what would be considered 'remanufacturing' in other sectors.

Given the potential for confusion, the 2018 IRP Report [1] distinguished between each of the VRPs, and adopted VRP definitions and terminologies that are consistent with internationally recognized sources (where they exist) that include, but are not limited to, the Basel Convention Glossary of Terms (Document UNEP/CHW.13/4/Add.2) [7] and Directive 2008/98/EC [8] (Figure 1.1).

1.2.1.1 Arranging Direct Reuse

Arranging direct reuse refers to: "*The collection, inspection and testing, cleaning, and redistribution of a product back into*

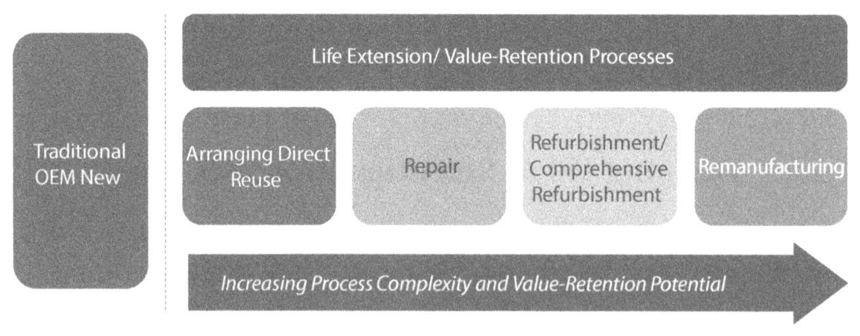

Figure 1.1 Definitions and structure of value-retention processes within this report.

the market under controlled conditions (e.g. a formal business undertaking)" [7] (Figure 1.2). Arranging direct reuse does not include reuse that occurs mostly through the undocumented transfer of a product from one consumer to another. Under arranging direct reuse, no disassembly, removal of parts, or addition of parts occurs. Only those products that are in sufficient working condition, not requiring any component replacement or repair, and to which quick and easy aesthetic touch-ups can be performed, qualify as arranging direct reuse products. These products are not guaranteed to meet original specifications and are typically offered to the market at a significant price discount, with no, or at least a much-modified, product warranty.

Arranging direct reuse becomes possible when a product reaches the end of its useful service life prematurely: the owner may require an upgraded product, may no longer need the product, or may have a change in preferences. Alternately, the usage/service requirement rate may have been less than expected during the products service life. In any case, although the product has reached end-of-use (EOU), it has not yet fulfilled its expected life or potential life. Arranging direct reuse enables the product to continue to maintain productivity through use, instead of being prematurely discarded into a waste or recycling system.

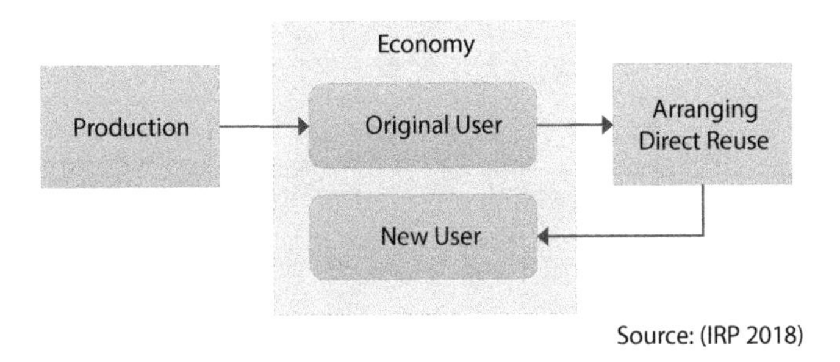

Figure 1.2 Descriptive summary of arranging direct reuse process.

1.2.1.2 Repair

Repair refers to: "*The fixing of a specified fault in an object that is a waste or a product and/or replacing defective components, in order to make the waste or product a fully functional product to be used for its originally intended purpose*" [7] (Figure 1.3). Repair activities include those required for known product issues, which enable the product to complete its original expected life. They also include the maintenance of a product that if left unmaintained, would have a constrained service life and/or utility.

Repair activities are performed at the product-level: an otherwise functional product must have some worn or damaged parts removed and new parts added for it to continue functioning for the duration of its expected life. Rather than the entire product being discarded into a waste or recycling stream due to a worn or damaged part, repair activities bring the entire product back to its original functioning capacity for the continuation of the product's expected life.

1.2.1.3 Refurbishment & Comprehensive Refurbishment

There are differing degrees of refurbishment activity that yield differing levels of material value retention and product utility: Refurbishment and Comprehensive Refurbishment (Figure 1.4).

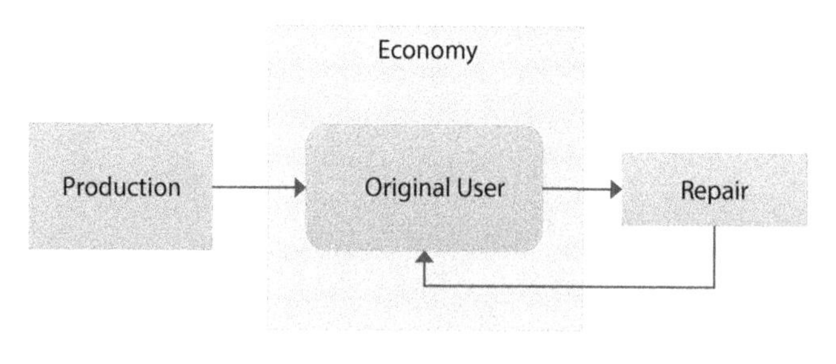

Figure 1.3 Descriptive summary of repair process.

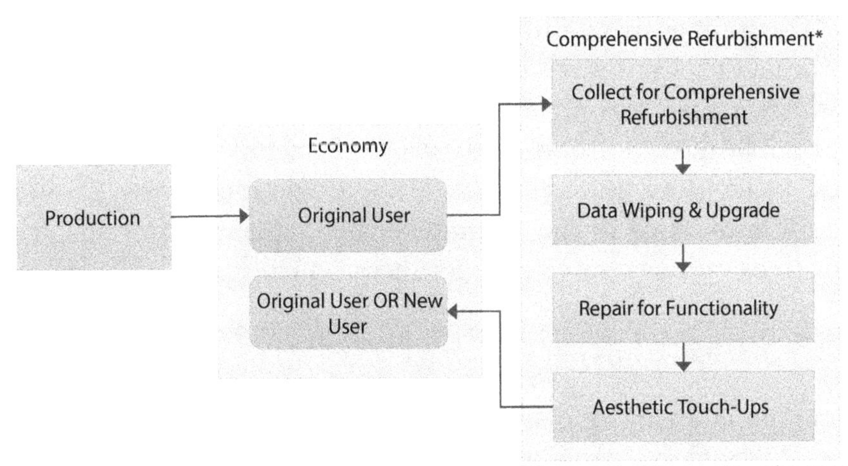

Figure 1.4 Descriptive summary of comprehensive refurbishment process.

- Refurbishment refers to: "*The modification of an object that is waste or a product to increase or restore its performance and/or functionality or to meet applicable technical standards or regulatory requirements, with the result of making a fully functional product to be used for a purpose that is at least the one that was originally intended.*" [7]
- Comprehensive Refurbishment refers to: "*Refurbishment that takes place within industrial or factory settings, with a high standard and level of refurbishment.*" [1]

Standard "Refurbishment" refers to 'minor overhauls' (heavy-duty engines and equipment), and 'upgrades' (electrical and electronic equipment). The refurbishment process is performed *within repair and/or maintenance facilities* to increase or restore performance and/or functionality or to meet applicable technical standards or regulatory requirements.

"Comprehensive Refurbishment" differs from standard refurbishment in that it involves a more rigorous process within a factory setting, and is only undertaken by certain sectors including, but not limited to industrial digital printers, medical equipment, and heavy-duty and off-road (HDOR) equipment parts. The addition of value during comprehensive refurbishment enables an almost full new service life for the product.

1.2.1.4 Remanufacturing

Remanufacturing refers to: "*A standardized industrial process that takes place within industrial or factory settings, in which cores are restored to original as-new condition and performance or better. The remanufacturing process is in line with specific technical specifications, including engineering, quality, and testing standards, and typically yields fully warranted products. Firms that provide remanufacturing services to restore used goods to original working condition are considered producers of remanufactured goods*" [1] (Figure 1.5).

This definition aligns with others from the literature, including from the WTO [2], Nasr and Thurston [5], USITC [3], and six global automotive remanufacturing associations[1] [6].

For a VRP to be considered 'remanufacturing', there is a minimum expectation of an industrial process in an industrial setting, consisting of specific activities including disassembly and cleaning, the requirement for testing and documentation, and the assurance of 'as-new or better-than-new' performance and quality of the remanufactured product.

---

[1] European Association of Automotive Suppliers (CLEPA), and European Organization for the Engine Remanufacture (FIRM), Motor & Equipment Remanufacturers Association (MERA), and Automotive Parts Remanufacturers Association (APRA), Automotive Parts Remanufacturers National Association (ANRAP), Remanufacture Committee of China Association of Automobile Manufactures (VRPRA).

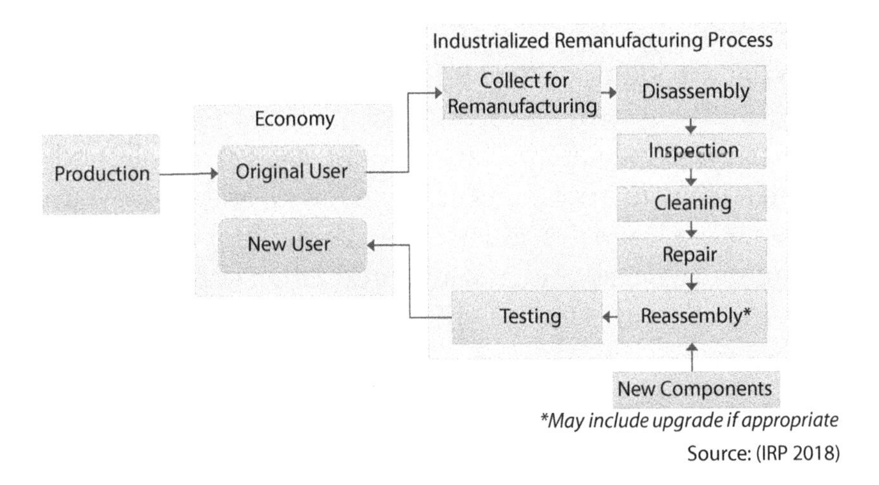

Figure 1.5 Descriptive summary of remanufacturing process.

The exact process undertaken by remanufacturers necessarily differs by product type: In most cases, remanufacturing includes the complete disassembly of all component parts for inspection and cleaning, however in the case of some products (e.g. industrial digital printers), disassembly only down to the module-level may be appropriate. This is especially true when the module itself has been designed for remanufacturing, in which case, by design, the module may have a longer expected service life than the product into which it is incorporated. Similarly, different sectors may utilize different reassembly procedures: In the case of medical devices, every disassembled part has an identifying serial number, and must be reassembled into the same remanufactured product; this differs from other sectors where disassembled parts may go directly into a general inventory and utilized as needed in the remanufacturing of completely different product units.

1.2.2 Expanded Systems-Perspective for VRPs

One of the most significant challenges to increasing the scale of VRPs in economies around the world is the complex nature of the circular economic system, which must consider stakeholders and conditions beyond those required under the traditional linear system. These include: collection infrastructure and incentives regulatory classifications and terminology that can interfere with access and trade; markets and social norms that associate 'new' products with status and quality; and well-entrenched technological and production systems oriented towards linear flows and heavily-scoped producer responsibility (Figure 1.6).

1.2.3 Evaluating the Value-Retention Potential of VRPs

In the context of VRPs, end-of-use (EOU) must be differentiated from end-of-life (EOL), as these critical terms clarify where opportunity for VRPs exist. In the design of new products, specifications for expected service life of the product are established. The expected life determines the designed durability and duration of the product: how many cycles, runs, miles, hours, etc. it should perform before maintenance interventions are required to ensure performance (e.g. repair, refurbishment), and how many of these can be performed before the product will degrade beyond use or reach EOL. Product EOL signifies that there are no other options for the product, but to be recycled or disposed into the environment. However, if other options exist to keep the product, and/or its components, within the market – via VRPs – then the product has only reached EOU.

The opportunity for VRPs lies in determining and understanding how a seeming product *EOL* may actually only be product *EOU*. In other words, once a product or components has reached EOU, it may be directed into EOL options of recycling or disposal – it may also, where infrastructure exists, be directed into a secondary market for repair, arranging direct reuse, refurbishment or comprehensive refurbishment, or remanufacturing instead.

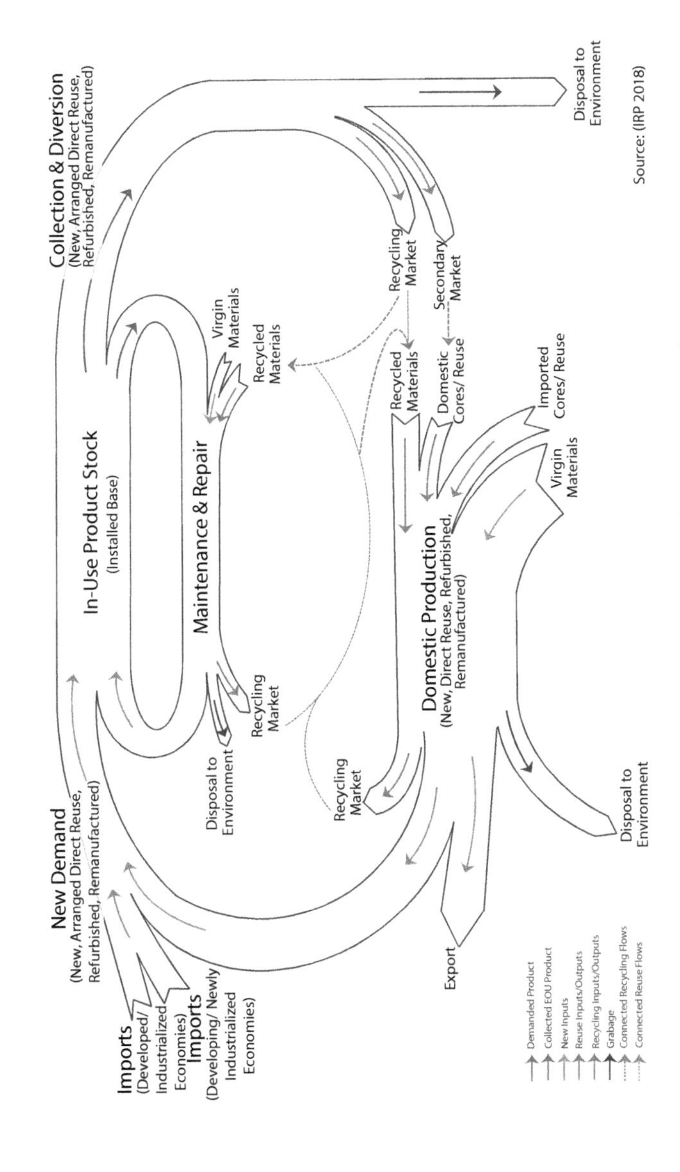

Figure 1.6 Description of the complex economic system required to support value-retention processes.

The IRP [1] found that VRPs could be organized into two categories (Figure 1.7):

- *Full Service Life Processes* refer to processes that enable the fulfillment of a complete new life for every usage cycle of the product, and includes manufacturing (Original Equipment Manufacturing, or OEM New), comprehensive refurbishment, and remanufacturing processes. These processes take place within factory settings and industrial operations;
- *Partial Service Life Processes* refer to processes that enable the completion of, and/or slight extension of, the expected product service life, through arranging direct reuse of the product, repair, and refurbishment. These processes take place within maintenance or intermediate maintenance operations.

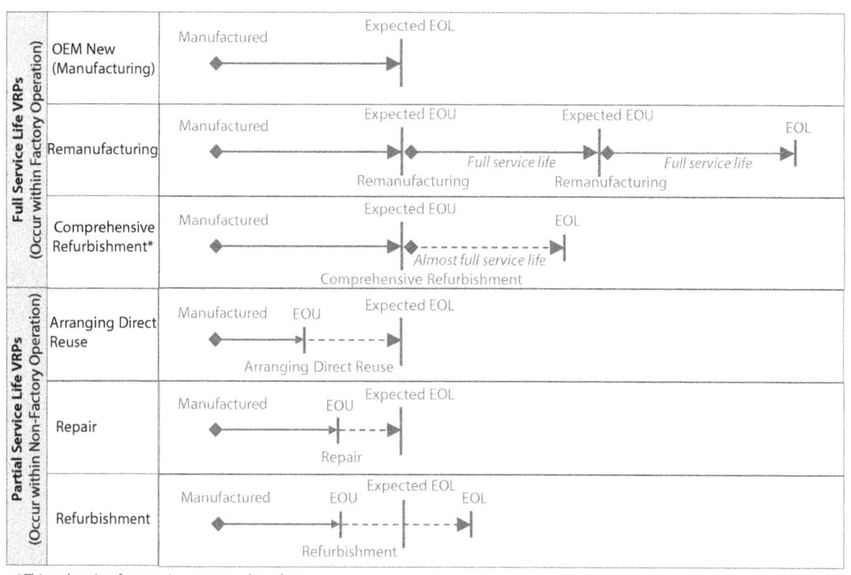

Figure 1.7 Summary of value-retention processes differentiation within the context of end-of-use (EOU) and end-of-life (EOL).

1.3 Value-Retention Process Evaluation Results

Although is it common to consider and discuss VRPs as 'equivalent' under a broad terminology of 'reuse', to do so would be problematic and misrepresentative. This is because each VRP is distinct in how it affects the product lifecycle, retains material and embodied value, and generates utility for the user.

1.3.1 Environmental Impacts of Value-Retention Processes at the Product-Level

In its assessment of VRPs, the IRP was interested in the environmental impacts of each VRP across five environmental metrics:

1) New material requirement (kg);
2) Embodied energy (MJ) associated with the extraction and processing of raw materials prior to production;
3) Embodied emissions (kg CO_2-eq.) associated with the extraction and processing of raw materials prior to production;
4) Process energy (MJ) required for gate-to-gate production activity;
5) Process emissions (kg CO_2-eq.) generated as a result of gate-to-gate production activity.

Based on IRP case study product research and analysis, the material efficiency, energy requirement, and emissions generation associated with US-based production of case study products, by OEM New and VRPs are presented in Figure 1.8 through Figure 1.10 [1].

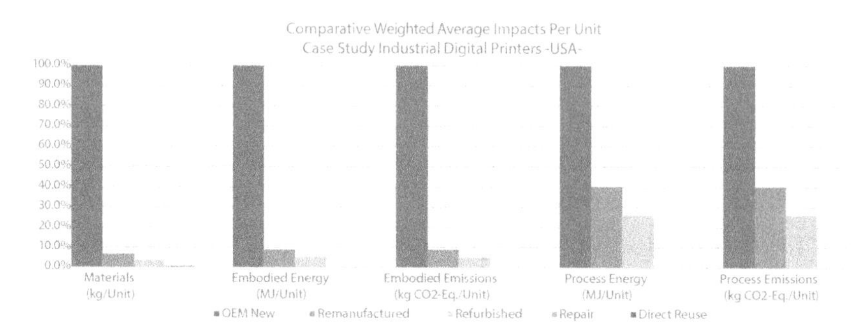

Figure 1.8 Impact reduction potential for USA via value-retention processes for industrial digital printers.

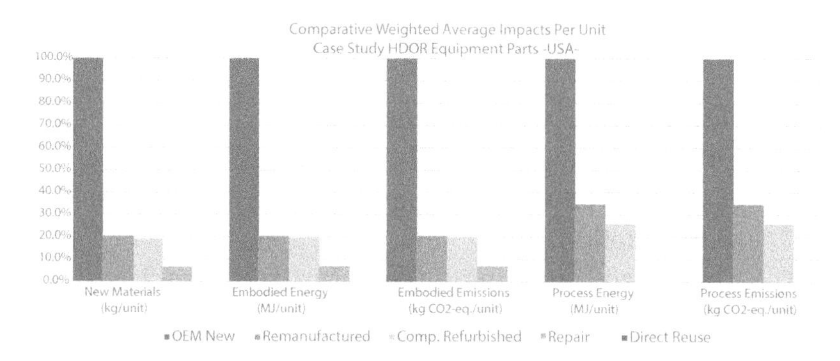

Figure 1.9 Impact reduction potential for USA via value-retention processes for HDOR parts production.

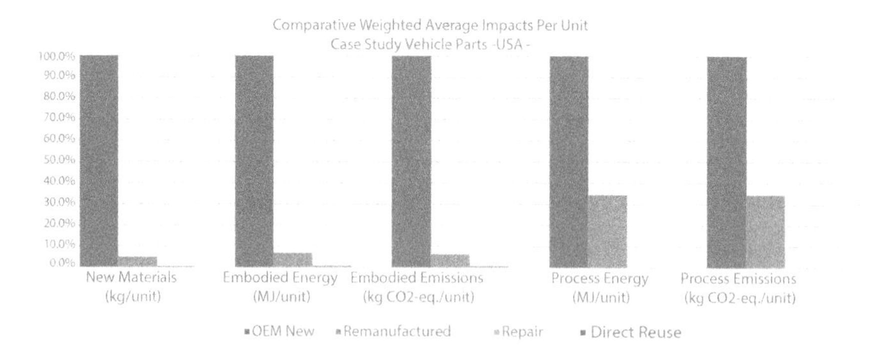

Figure 1.10 Impact reduction potential for USA via value-retention processes for vehicle parts production.

1.3.2 Economic Advantages of Value-Retention Processes at the Product-Level

The economic considerations of VRPs at the product level are also highly relevant to the discussion of impacts and benefits that become possible through the use of VRPs in the pursuit of circular economy. In its study the IRP assessed the select case study products, presented in Figure 1.11 through Figure 1.13, across the following economic metrics:

1) Production waste (including scrap recyclable material) (kg/unit);
2) Cost advantage (% $USD per unit vs. OEM New); and
3) Employment opportunity (Full-time Laborer or FTE/unit).

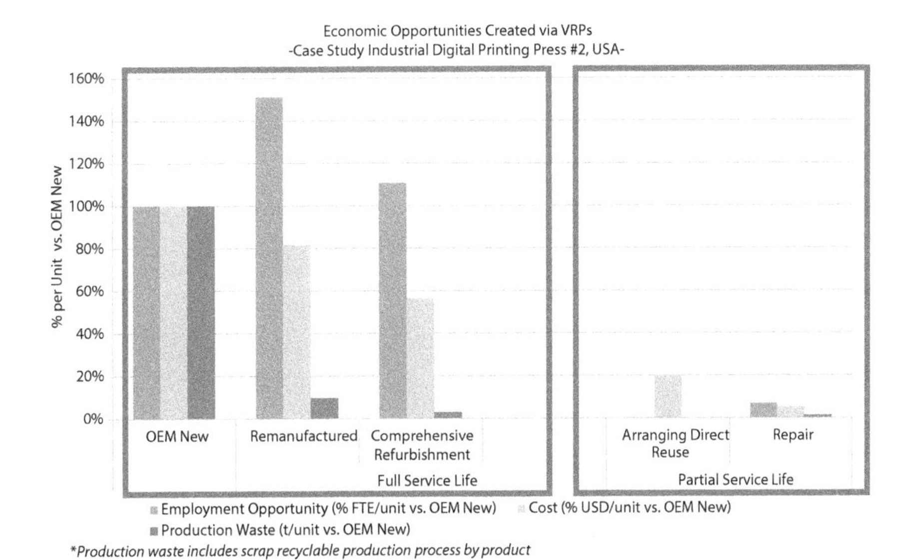

Figure 1.11 Employment opportunity, cost advantage, and production waste reduction via VRPs for case study industrial digital printers.

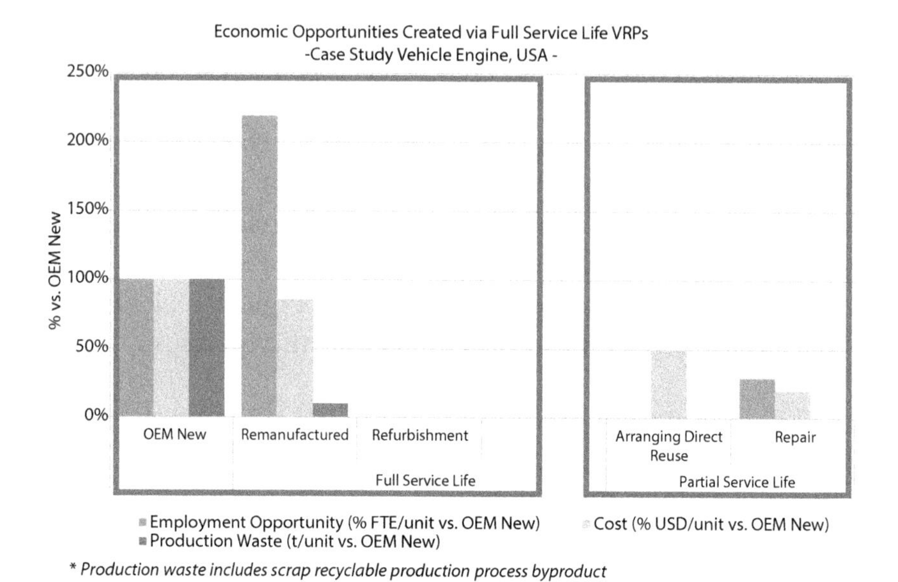

Figure 1.12 Employment opportunity, cost advantage, and production waste reduction via VRPs for case study vehicle parts.

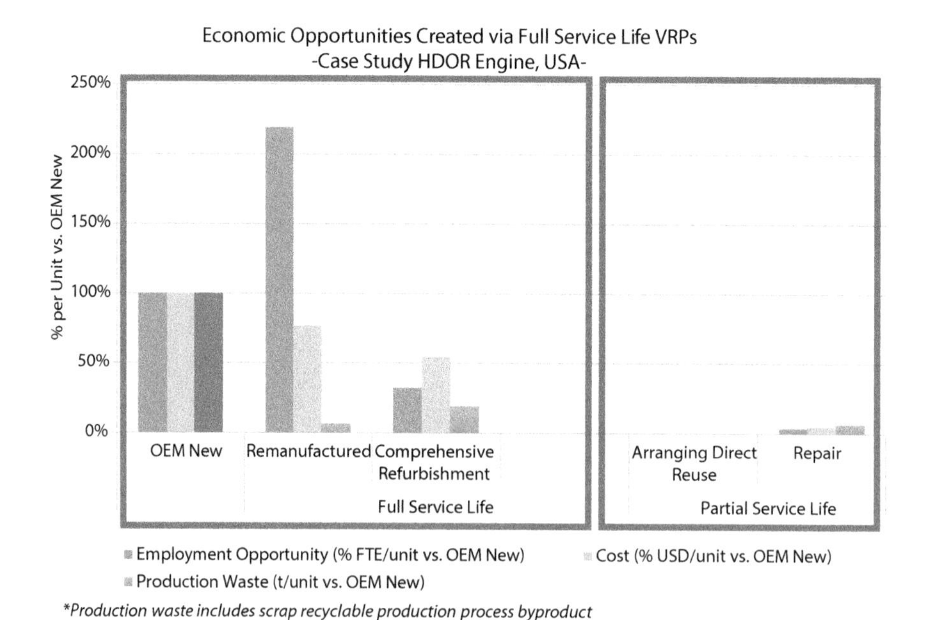

Figure 1.13 Employment opportunity, cost advantage, and production waste reduction via VRPs for case study HDOR equipment parts.

1.3.2.1 *Production Waste Reduction through Value-Retention Processes*

Through the reuse of viable products/components, significant processing activities are no longer required, and the waste associated with those processes is offset. Every VRP offers some degree of reduced production waste: where arranging direct reuse requires no new material inputs, and therefore no additional production wastes, remanufacturing enables a production waste reduction between 90% (industrial digital printers) and 95% (vehicle parts). The decrease in the volume of production waste and recyclables is first and foremost an economic opportunity associated with increased adoption of VRPs: this reduction in waste leads to reductions in storage, hauling and tipping fees that would otherwise be borne by the producer.

1.3.2.2 *Production Cost Advantages of Value-Retention Processes*

Significant cost advantages are made possible through VRPs, as a large share of costs to the producer are offset by the reduced requirement for new input materials and associated processing costs. In addition, for some products and sectors, process energy-related costs can be significantly reduced by offsetting many processing stages and activities with more manual activities. According to the IRP Report [1], cost advantages of VRPs range, conservatively, between 15% and 80% of the cost of an OEM New version of the product.

1.3.2.3 *Employment Opportunities through Value-Retention Processes*

There is significant employment opportunity inherent in VRPs. While the cost of labor remains a significant share of total production costs in all manufacturing activity, in the case of VRPs, the additional labor cost is typically more than offset by the relative reduction in materials, utilities, and other overhead and operating costs. In the case of remanufactured products, a significant increase in full-time labor requirement is observed, and at the same time, remanufacturers are typically able to offer a consistent cost advantage to potential customers. It is important to note that the employment opportunity is not equal across all VRPs: Only remanufacturing, and to some degree comprehensive refurbishment, offer greater full-time employment opportunity than traditional OEM New production. Thus, as the production share of remanufacturing and refurbishment are increased, a corresponding increase in full-time employment opportunities is possible.

1.3.3 Systemic Barriers to VRPs

The objective of increasing the scale and prevalence of VRPs and products within an economy requires a holistic approach that considers the magnitude and cause of barriers throughout the entire system, as well as how those barriers may interact to compound or negate one another. There are four main categories of barriers to VRPs according to the IRP, outlined below:

1. **Regulatory and Access Barriers**: Refers to barriers that restrict the movement of, and/or access to VRP products or cores. These barriers may manifest as prohibitions of the production and/or sale of VRP products into a domestic market; they may also manifest as increased fees, tariffs or other transactional costs associated with bringing finished VRP products or components (cores) for VRP production into the domestic economy.
2. **Collection Barriers**: VRP production is dependent on the ability to access EOU products and components; the majority of economic and environmental benefits created via VRPs are tied to the offset or original production materials and processes through the reuse of viable 'cores'. If collection

infrastructure is inadequate or inefficient, the reuse input requirements of VRP producers cannot be met.

3. **Technological Barriers**: When technology, product knowledge, process know-how and/or skilled labor are insufficient, the capacity of the VRP producer is relatively constrained, and the associated potential economic and environmental benefits are limited. In such cases, in addition to being limited in the current state, the VRP producer's ability to build capacity over-time – whether demand opportunity exists or not – is likely stunted.
4. **Market Barriers**: The complexity of customer (consumer) attitudes, preferences, willingness-to-pay, and actual purchasing behavior creates significant additional challenges for VRPs, even in markets where no other barriers are present. Where a strategic approach for many VRP producers is to offer a discounted price to incentivize the purchase of the VRP product, this price discount may also limit the VRP producer's ability to maximize profit margins and find cost advantage in the production process.

1.4 Key Insights Regarding VRPs

1.4.1 Value-Retention Processes Create Net-Positive Outcomes for Circular Economy

At the product-level, offset embodied energy and emissions create immediate and obvious ranges of potential impact reduction and value-retention associated with the adoption of VRPs. When considered in the context of the process definitions, and the subsequent quality and performance of the VRP product, it becomes clear that different VRPs are appropriate for different objectives. Where remanufacturing and comprehensive refurbishment (Full Service Life Processes) add and retain relatively greater value in the system in terms of materials and functional form, for some products and economies these relatively more intensive industrial processes may increase the associated process energy requirement and process emissions. At the same time, the rigorous industrial process can lead to greater economic opportunities in the form of increased labor requirement, decreased waste management costs, and greater utility, for the customer.

In contrast, arranging direct reuse, repair, and refurbishment (Partial Service Life Processes) can be undertaken at a relatively lower cost, enabling customers with budget constraints to continue participating in the market; and they can be completed with lesser material requirement, energy requirement and associated emissions and waste. However, Partial Service Life processes offer only limited value and utility to the customer and retain less value in the system over time.

1.4.2 Product-Level Efficiency Gains Lead to Economy-Level Efficiency Gains

There is often a perception that the pursuit of sustainability must come at an economic cost. While this perception may be warranted in a short-term context, through the adoption of VRPs, significant impact reduction can be achieved at the same time that economic opportunity is being created. The reduction in new material input requirement, and the embodied value inherent in the already-functional form, ensure that VRPs can offset a significant share of costs that would otherwise be required for OEM New production. This cost advantage to the producer generates additional economic opportunities in several ways: first, with lower operating costs there are fewer cost barriers to entry into the marketplace for potential VRP producers, and this can support and enable faster scale-up within domestic industry; and second, lower operating costs may enable VRP producers to pass the cost advantage along to their customers.

Lower-priced VRP product options in the market can enable new segments of customers to participate where budget constraints may previously have prevented such engagement.

The relative positive outcome of avoided impacts can be observed across each sector and economy, and highlights the importance of utilizing a systems-view when assessing the potential for VRPs within the circular economy:

1. Any increase in VRP production reduces average new material demand and creates an opportunity to avoid requirement for new materials.
2. The avoidance of new material inputs creates significant benefits in avoided embodied materials energy and embodied material emissions impacts that would otherwise be incurred through the extraction and primary processing of those new input materials. Regardless of which VRP is adopted, a net-positive reduction in embodied energy and embodied emissions is consistently observed across every sector and economy.
3. The significant retention of the functional form and material value of component parts enabled via VRPs offsets the production waste associated with original production.

1.4.3 The Mechanics of a System Designed for Value-Retention Processes

There are many existing attributes and aspects of current production systems that can be leveraged in the pursuit of a system designed for optimized VRP production. While every economy faces different challenges and barriers to VRPs, each also has an already established relationship with the key aspects of the VRP system that can inform a policy and implementation strategy:

- For economies that currently engage in diversion and collection to recycling markets, these systems can be adapted to include diversion to secondary markets for reuse and VRP production;
- For economies that do not engage in collection or reverse-logistics, expertise in current forward-logistics systems can be leveraged to improve overall logistics system utilization and productivity, alongside the application of Best Practices that may have already been established for collection programs in other jurisdictions;
- For economies with technological barriers affecting producer capacity, the learnings about technology transfer enabled through improved access and trade in other products categories can be employed to the benefit of VRP production.

1.4.3.1 Value-Retention Processes are a Gateway to Recycling

There is a common misperception that VRPs may detract from, or compete against recycling; in fact, all VRPs and recycling are complementary and essential within the context of a circular economy. A hierarchical perspective on value-retention is useful: where *VRPs* ensure that both material value *and* functionality are retained within the product, once functionality has degraded it is the *recycling* system that ensures material value is still retained within the broader system. An alternate way of considering this complementarity is to consider that all products will eventually reach a point at which they no longer qualify for arranging direct reuse, repair, refurbishment or remanufacturing – either because of the associated cost, or because the implicit quality and utility-potential has been degraded. At that point, there is still an essential need for efficient and effective recycling systems to recover the value of the materials contained within the product, and to recirculate those materials back into circular materials economy.

1.4.4 Overcoming Barriers to Value-Retention Processes

1.4.4.1 Economic Conditions and Access to VRP Products

The presence of access barriers dominates the ability of an economy to realize the benefits from VRPs through uptake and diffusion. In the absence of clear domestic market demand and access to VRP technologies, potential VRP producers are unable to support the business-case for VRPs, in spite of the known environmental and economic benefits. Access barriers slow the growth of VRP production within the domestic economy, as well as the speed of VRP capacity scale-up, and the related growth in domestic demand for VRP products. Technology and knowledge transfer that is essential for enhancing the learning curve of domestic producers is inhibited, ultimately preventing opportunities for improved production and operational efficiency.

1.4.4.2 Market Challenges

Education and information is an essential strategy for combatting market barriers. A primary concern of customers in both open and restricted economies is the perceived quality discount associated with non-new products. There is a wide range of misinformation and mistrust related to VRP products. The lack of standards, certifications, and the prevalence of 'bad experience' stories have ensured that customers approach VRP products with high risk-aversion tendencies. Educating customers about the differences between VRPs (e.g. refurbishing vs. remanufacturing) so that *perception*s about quality can more appropriately reflect *actual* quality is an important strategy in developed/industrialized markets, which need to focus on growing demand for VRP products as viable alternatives.

Both industry and government have a role to play in overcoming market-based barriers to VRPs. Where government must necessarily be concerned about the assurance of consumer interests and safety, industry has an opportunity to respond to these concerns through the development of standards, certification and additional merit-based systems that can offer reassurance to both customers and policy-makers.

1.4.4.3 Regulatory and Policy Opportunities

Comprehensive restrictions and prohibitions, and/or burdensome requirements debilitate natural market forces, and ensure that many potential customers are unable to participate in the current market due to access and/or budget constraints. It is possible to ensure the protection of consumer safety and interests without comprehensive prohibitions and restrictions. The development and employment of VRP standards, certifications, and other non-regulatory forms of enforcement are being investigated and implemented currently in developed/industrialized economies.

1.4.4.4 Diversion & Collection Infrastructure

VRPs are reliant on the diversion and collection of EOU products for use as inputs to the process; while individual companies may have established their own networks and collection infrastructure to ensure sufficient supply of reuse inputs, this can create a significant and inefficient cost-burden on the individual organization. Other examples of shared collection infrastructure, such as e-waste diversion and packaging extended producer responsibility (EPR) programs have demonstrated the ability to both increase collection rates and distribute the costs of operating the system. The requirement that these systems be funded by industry can provide an incentive to pursue greater cost-efficiency and performance over time. Creative and shared approaches to collection infrastructure may have some merit in cases where the objective is to increase collection and retention of value within a system.

1.4.4.5 *The Nature of Barriers Must Guide Strategic Barrier Alleviation*

Economies face distinct combinations of VRP barriers and may have unique objectives for VRPs as part of an economic or environmental agenda. As such, there may be a range of potential strategic interventions available to policy- and decision-makers depending on each unique situation, as demonstrated in Figure 1.14.

1.5 Conclusions

The circular economy sets out a framework in which VRPs work alongside other essential economic and behavioral strategies to reduce the environmental burden of the global economy. However, this is a grand vision, and significant market transformation is required to achieve the potential economic and environmental benefits promised by circular economy.

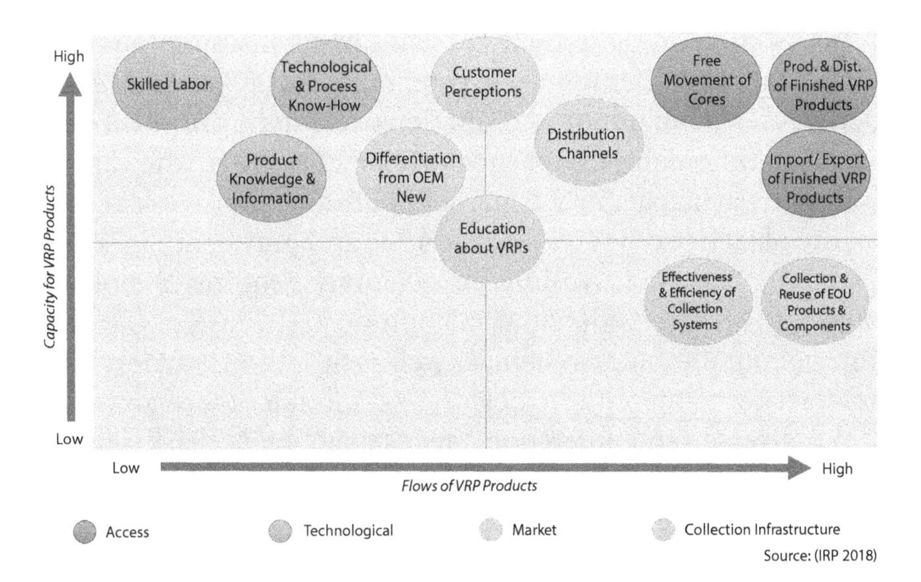

Figure 1.14 Differentiated barrier alleviation strategies for different economic objectives.

Ultimately, responsibility for scale-up and adoption rests with every decision-maker on the planet: from the individual consumer making an everyday purchase decision, through to the policy-maker considering how to plan for economic growth in the context of international GHG emissions reduction commitments. The next market transformation must ensure a shift in understanding, awareness, access and adoption of VRPs for each one of these decision-makers.

There are a range of opportunities to help kick-start this much-needed market transformation, and both governments and industry have an important role to play in helping to increase the adoption of VRPs. The inclusion of VRPs within the domestic production mix has been shown to create net-positive per-unit reduction in new material requirement, embodied material energy, embodied material emissions, and in many cases, process energy and emissions as well. The increased presence of VRPs within an economy leads to an increase in avoided production impacts in every case. The most significant value-retention and impact avoidance comes from Full Service Life VRPs of remanufacturing and comprehensive refurbishment, despite their current low intensity in the most well-established economies. Arranging direct reuse and repair activities offer an important function of extending product utility at a minimal impact, particularly for customers who cannot afford to participate in the market in other ways.

The objective of increasing the scale and prevalence of VRPs and VRP products within an economy requires a holistic approach that considers the magnitude and cause of barriers throughout the entire system, as well as how those barriers may interact to compound or negate one another. There are multiple, diverse, and interconnected stakeholders, each with a potential role to play in the transition to circular economy and the uptake of VRP production:

- Government policy-makers have a central and pivotal role related to the presence and alleviation

of regulatory, access and collection infrastructure barriers;
- Other stakeholders, including industry, may have an important role to play in the alleviation of barriers related to the customer market and technological capacity.

For economies wishing to pursue circular economy and VRPs as a key aspect of an effective system, acknowledgement of the underlying order within the system can help to guide strategic policy opportunities, as simplified in Figure 1.15.

A simplified approach to barriers assessment and the role of government and industry members in developing strategic responses to barrier alleviation is outlined in Figure 1.16.

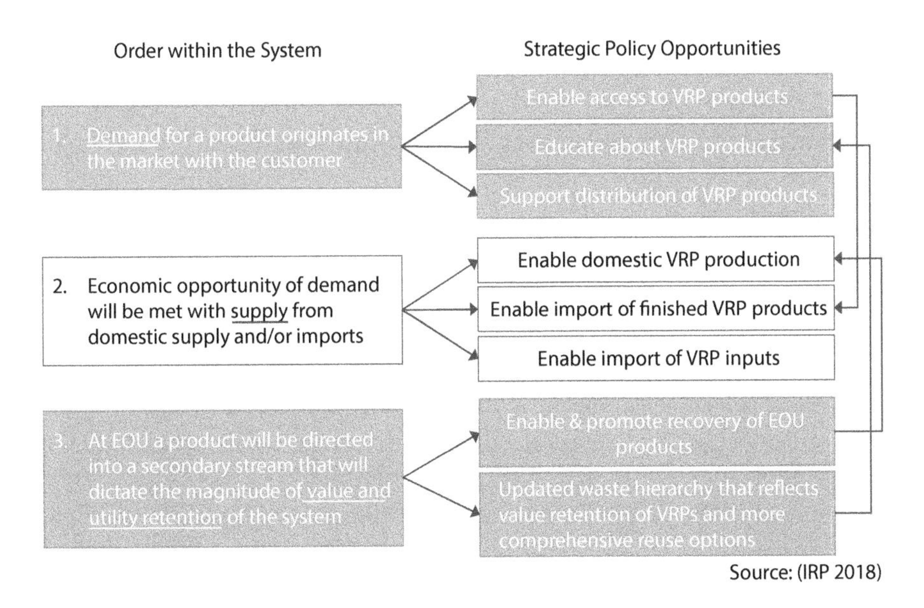

Figure 1.15 Inherent system order enables priorities for alleviation of VRP barriers.

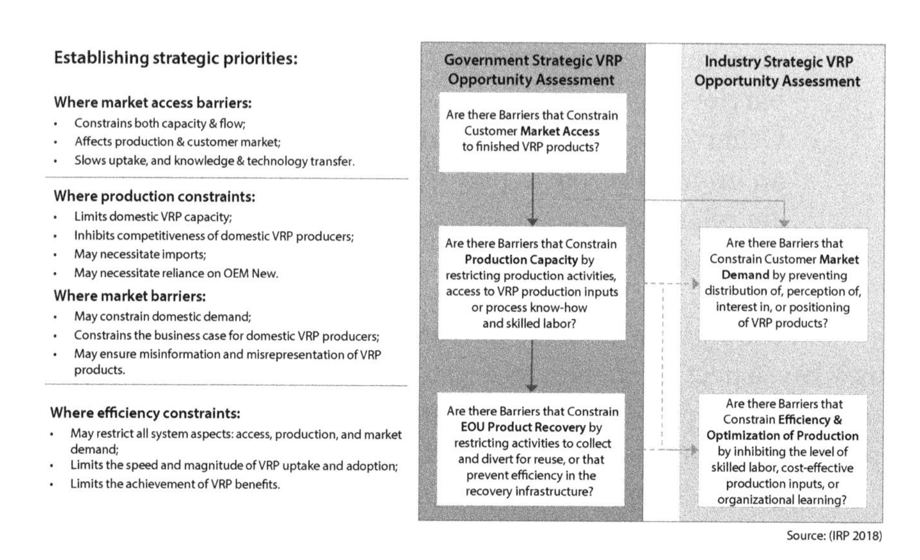

Figure 1.16 Role of government and industry decision-makers in assessment of VRP barriers and strategic priorities.

References

1. IRP, *Re-Defining Value - The Manufacturing Revolution: Remanufacturing, Refurbishment, Repair and Direct Reuse in the Circular Economy.* Edited by Nabil Nasr, Jennifer Russell, Stefan Bringezu, Stefanie Hellweg, Brian Hilton, Cory Kreiss and Nadia von Gries, United Nations Environment Programme, Nairobi, Kenya, 2018.
2. World Trade Organization, *Market Access for Non-Agricultural Goods: Answers to Questions from China on Remanufacturing.* WTO Document TN/MA/W/122, Geneva, 2009.
3. U.S. International Trade Commission, *Remanufactured Goods: An Overview of the U.S. and Global Industries, Markets and Trade*, U.S. International Trade Commission, Washington, D.C.: U.S., 2012.
4. Hardin, W. and Stone, M., Remanufacturing of Electrical Equipment vs. Reconditioning or Rebuilding of Electrical Equipment. *Professional Electronic Apparatus Recyclers League*, 2012. Accessed from http://www.pearl1.org/PEARL-News/remanufacturing-reconditioning-electrical-equipment.htm.

5. Nasr, N. and Thurston, M., Remanufacturing: A key enabler to sustainable product systems. Proceedings of the 13th CIRP International Conference on Lifecycle Engineering, Leuven, 2006.
6. Motor & Equipment Remanufacturing Association, *Remanufacturing Associations Agree on International Industry Definition [Press Release]*, accessed 15 Sep 2016. https://www.mera.org/news/remanufacturing-associations-agree-international-industry-definition-0, 2016.
7. Conference of the Parties to the Basel Convention on the Control of Transboundary Movements of Hazardous Wastes and Their Disposal. Technical guidelines on transboundary movements of electrical and electronic waste and used electrical and electronic equipment, in particular regarding the distinction between waste and non-waste under the Basel Convention, Appendix 2: Glossary of Terms. edited by COP 13: United Nations Environment Programme, 2017.
8. European Commission, Directive 2008/98/EC of the European Parliament and of the Council of 19 November 2008 on waste. *Off. J. Eur. Union L*, 312, 13, 22.11, 2008.

2

The Role of Remanufacturing in a Circular Economy

Erik Sundin

Division of Manufacturing Engineering, Department of Management and Engineering, Linköping University, Sweden

Abstract

The circular economy is a concept that is studied by many manufacturers today, especially in Europe. Remanufacturing is increasingly recognized as an essential part of the circular economy. This chapter describes the characteristics of both the circular economy and remanufacturing, as well as the relationship between the two, and their mutually reciprocal business and ecological benefits. Evidence from research and industry supports the notion that a strong connection exists between the adoption of remanufacturing and the success of the circular economy.

Keywords: Remanufacturing, circular economy, closed-loop system, lifecycle, business models, product service system, sustainable, environment

2.1 Introduction

From resource constraints to evolving customer demands, manufacturing companies face many challenges in the present competitive market. Remanufacturing is one of many ways to stay

Email: erik.sundin@liu.se

Nabil Nasr (ed.) Remanufacturing in the Circular Economy (31–60)

competitive, bringing business benefits such as reduced costs and increased market shares while simultaneously improving environmental performance. Concurrent to the emergence of remanufacturing in today's industrial systems, the concept of a circular economy has gained popularity in recent years, especially amongst authorities within Europe, as a means to achieve the critical decoupling of economic growth from environmental degradation. Remanufacturing is often seen as a central part of this circularity; this chapter aims to describe its role and explore why its adoption is critical to the success of a circular economy.

2.2 Circular Economy

2.2.1 What Is It?

The concept of the circular economy is described in many ways by different authors. The Ellen MacArthur Foundation, a leading advocate of the circular economy, merges these, describing the circular economy as an industrial system "restorative and regenerative by design, aiming to keep products, components, and materials at their highest utility and value at all times.[1]"

One way to illustrate this concept, as conceived by the company Circle Economy, is illustrated in Figure 2.1. This diagram helps us visualize the conventionally linear manufacturing economy and how interventions at the latter stages of a product's lifecycle may circularize this system. This "snail" diagram focuses on the shift from linear to circular by closing loops at the end of the traditional chain in order to create a zero waste system in which the embedded value and complexity of products is preserved. This preservation relies first on the optimization of use, and subsequently (in descending order of economic and

[1] www.ellenmacarthurfoundation.org/circular-economy/.

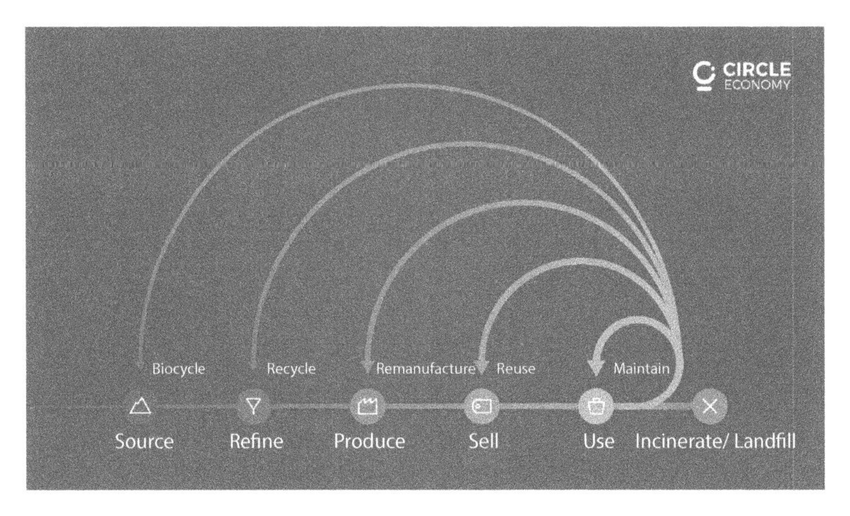

Figure 2.1 Circular economy illustrated as a "snail", created by the Circle Economy company.

environmental preferability) several post-use lifecycle stages—maintenance, reuse, remanufacturing, and material recycling—that continue to capture a product's intrinsic value rather than relinquish it through landfill disposal.

While the snail diagram emphasizes the emergence of the circular *from* the linear, the Ellen MacArthur Foundation's (EMF) "butterfly" diagram[2] illustrates that in reality, relationships between even upstream lifecycle stages are necessarily more complex in order to create stability—as in the ecological concept of a food *web* rather than a singular chain (Figure 2.2).

In this conceptual structure, the circular economy consists of two major flows of materials: *biological material* and *technical material*. The EMF view suggests that materials should be kept within circular loops from extraction and throughout their lifecycles, while minimizing the material flowing to energy recovery (incineration) and landfilling as absolute last resort options. Importantly, remanufacturing is highlighted

[2] www.ellenmacarthurfoundation.org/circular-economy/interactive-diagram.

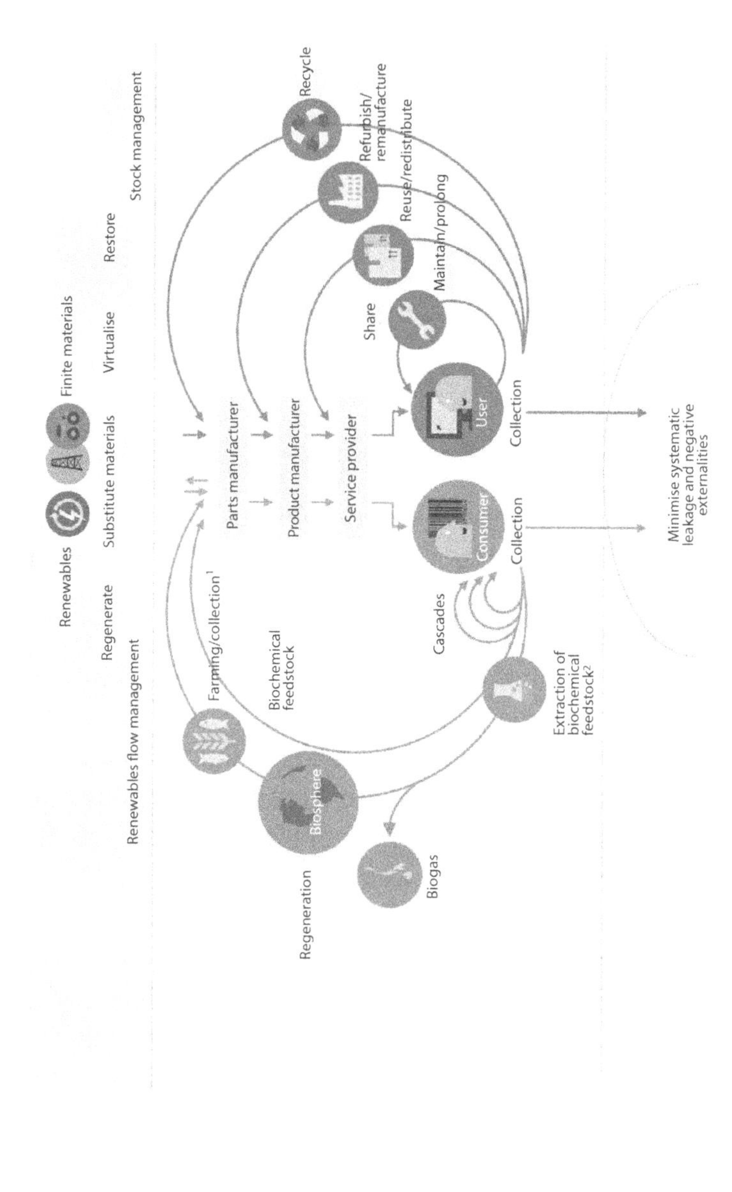

Figure 2.2 Circular economy illustrated as a "butterfly", created by the Ellen MacArthur Foundation.

in the flow of technical materials (the right-hand wing of the butterfly) as one of the options for getting products, parts, and materials back from the use phase.

Beyond the EMF, the European Union has also been actively working to understand the circular economy concept and support a transition to such a sustainable system. Policy-driven research such as *Roadmap to a Resource Efficient Europe* (2011) and *Towards a circular economy: A zero waste program for Europe*[3] (2014) describes how moving towards a more circular economy is essential to achieve resource-efficient growth, and contends that sustained improvement of resource-efficient performance—through actions and technologies presently within our reach—can foster major economic benefits.

The fundamental concept underlying all of this is that circular economy systems minimize waste by keeping the added value in products for as long as possible. They keep resources within the economy when a product has reached the end of its use so that it can be used again repeatedly, continuing to create and hold value for its customers over multiple cycles of use. However, a transition to a more circular economy requires changes throughout the entire product value chain, from product design to business models. These must necessarily be supported by technologies and methodologies that enable success, from finding new ways of turning waste into resources to changes in the behavior of product users and consumers. This implies systemic changes on a large scale, and innovation not only in technologies, but also in organizations, social paradigms, and economic structures [1].

2.2.2 How Does It Work?

The authors of the EU report [1] illustrate the circular economy as a conceptual circular system with seven phases: (1) raw material,

---

[3] http://eur-lex.europa.eu/legal-content/EN/TXT/?uri=CELEX:52014DC0398.

(2) design, (3) manufacturing/remanufacturing, (4) distribution, (5) consumption/use/reuse/repair, (6) collection, and (7) recycling. These phases are interlinked, as materials can be used in a cascading manner and are not necessarily constrained to a single pathway based on their origin; for instance, industries may exchange by-products as repurposed outputs-to-inputs, products may be refurbished or remanufactured, or consumers may choose product service systems over ownership and disposal. The aim with these circular flows is to minimize the volume of resources escaping from the circle so that the system functions with optimal efficiency. The circular economy is then described as the actions and mechanisms occurring within these seven phases to keep value while avoiding waste.

To facilitate the transition to circular economy, the European Union suggests that innovations needed throughout the product value chain should, at a high level, include:

- reducing the quantity of materials required to deliver a given service (*lightweighting*)
- lengthening products' useful life (*durability*)
- reducing the use of energy and materials in the production and use phases (*efficiency*)
- reducing the use of materials that are hazardous or difficult to recycle (*substitution*)
- creating sustainable markets for secondary raw materials (*recyclates*)
- designing products that are easier to maintain, repair, upgrade, remanufacture or recycle (*ecodesign*)
- developing the necessary services and infrastructures for consumers in this regard
- incentivizing and supporting waste reduction and high-quality separation by consumers
- creating separation and collection systems that minimize costs of recycling and reuse

- clustering activities to prevent by-products from becoming waste (*industrial symbiosis*)
- encouraging wider and better consumer choice through renting, lending, or sharing services as an alternative to owning products, while safeguarding consumer interests (in terms of costs, protection, information, contract terms, insurance aspects, etc.).

In order to achieve a circular economy, a crucial starting point is the design of manufacturing processes, products, and services. Products must be redesigned specifically to enable them to be used longer, repaired, upgraded, remanufactured or eventually recycled, rather than discarded (usually in landfill) and replaced. Manufacturing processes must both enable reusability of products and, accordingly, accommodate the reuse of secondary raw materials from restorative sources in subsequent manufacturing cycles. Likewise, innovative business models must create a new relationship between companies and consumers, e.g. through product service systems in which a company sells the function of a product rather the actual product, shifting the source of value from the disposition and replacement of that product to the maintenance of that product's ability to function. This is done, for instance, when selling the function of imaging equipment by a number of copies, rather than a selling a physical photocopier. In this example, the customer only pays for the equipment's ability to function over a defined period, rather than the material and energy embodied in the machine itself and wasted when discarded [2].

Prior to the maturation of the circular economy concept, researchers used the general terms of *closed-loop systems* or *product lifecycle* to describe the notion that the impacts and value of the manufacturing economy exist—and must be addressed—well beyond the factory door. Across each discrete perspective on closed-loop systems and product lifecycles, it is clear that each is built around similar concepts through which the physical life of a product[4] can be divided into the three major phases—the beginning, middle and end of physical life—which encompass the entirety of intermediate lifecycle activity stages. In each, these include:

- **Beginning-of-Life (BoL):** material acquisition, design, manufacture of parts, assembly
- **Middle-of-Life (MoL):** product delivery, installation, use, repair, maintenance, refurbishment, remanufacturing, reuse (second through nth use), and final use
- **End-of-Life (EoL):** material recycling, upcycling, energy recovery, and landfilling

Note here that remanufacturing takes place at the product's middle-of-life and not at the product's end-of-life. This distinction stands on the grounds that a product's *physical life* is the duration of time for which it remains in the form and executes the function of its design. Products recovered for remanufacturing are still physically products, and although they have reached the end of their *usable* life, they are simply at the end of their use (EoU), not the actual EoL. Remanufacturing is therefore technically a method to extend the *physical* lifecycle of a product, although each remanufacturing cycle ostensibly creates an entirely new *usable* life. At the true EoL, however, the value of a product's form and function may either be exhausted or no longer economically recoverable, and thus products are reduced their constituent materials from which some value may still be extracted through material recycling and energy recovery, where as a last resort, waste can be incinerated and the resultant heat can used for electricity generation and heating.

[4] In business theory, a product's lifecycle often refers to a product's business life phases, namely *introduction*, *growth*, *maturity*, and *decline* (see e.g. Cox Jr WE. Product life cycles as marketing models. Journal of Business, 1967; 40:4). However, the product life cycles studied in this chapter are the physical ones described in Figure 2.3.

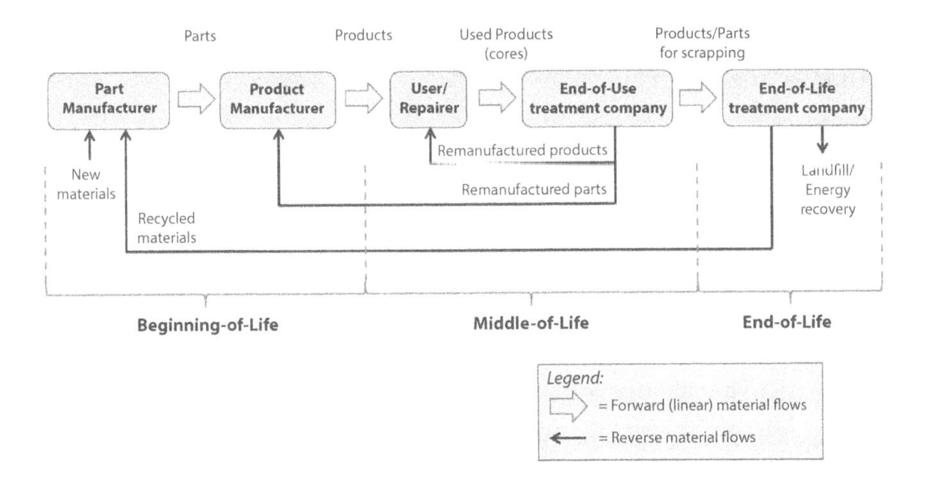

Figure 2.3 Major phases of the product lifecycle, including the forward and reverse flows of products, parts, and materials (adapted from [3]).

EU reports that existing infrastructure, business models, and technology—together with established producer and consumer behavior—keep economies locked in to the linear model, and thus increase the resistance to and cost of transitioning towards a circular economy[5]. In many cases, these barriers are rooted simply in the lack of information, confidence, and capacity to move to circular economy solutions. Further, the financial system as we know it today is built upon historic notions of economic growth tied to physical growth. As a result, it often fails to provide justification for or support the viability of investment innovative circular business models, because these—in their differentiation from the conventional—are (often erroneously) perceived as more risky and complex, deterring many traditional investors. Conventional consumer habits can also hinder new product and service development. Such barriers tend to persist in a context where prices do not reflect the real costs of resource use to society, and where policy fails to

[5] The EU report "Towards a circular economy: A zero waste programme for Europe" can be found at: http://eur-lex.europa.eu/legal-content/EN/TXT/?uri=CELEX:52014DC0398.

provide strong and consistent signals for the transition to a circular economy [1].

2.2.3 Summary

There are many ways of achieving circularity (i.e. reverse flows of products, parts and materials). Despite their particularities, each is connected by common themes: a preference for long lifecycles and short material loops that keeps resources and value in use and prevents waste generation. Upgradability and reconfigurability of products and parts is also preferable, when possible, to meet evolving customer demands over time.

Furthermore, the diverse options of circularity are each differently suitable for different products. For example, products like used bicycles and clothing are easy to pass directly from one user to another with no restorative intervention (reuse), while used toner cartridges and car engines require specific processes and equipment to restore functional value, and thus are better suited to remanufacturing. This multitude of available options for circularity, however, is beneficial to our equally diverse industrial economies; that is to say, the ability of different industry sectors to choose the method of circularity that best suits the characteristics of their industry indeed maximizes the opportunity for systemic circularity well beyond the possibility of assimilating all industrial economic process into a singular circular pathway.

2.3 Remanufacturing

2.3.1 What is Remanufacturing?

The term "remanufacturing" has been used since the late 1930s, or even earlier, and it has been defined differently by industrial practitioners and researchers across diverse backgrounds. Hence, there exist many definitions of the term "remanufacturing", but most are founded in a common basic idea of restoring a product's original function. In my research, I have used this general definition:

> *"Remanufacturing is an industrial process whereby used products (cores) are restored to useful life. During this process, the core passes through a number of remanufacturing operations, e.g. inspection, cleaning, disassembly, reprocessing, reassembly, and final testing, to ensure it meets the desired product standards"* (adapted from [4]).

Here, a core is defined as a discarded EoU product that is collected for remanufacturing. In some cases, such as electronics, product cores sent for remanufacturing may no longer function as designed; in these cases, the entire product is remanufactured to create a functioning replacement. In others, such as brake calipers or toner cartridges, the central function of a product core may still be intact, but inhibited by the wear of a consumable component; in these, only the worn and predictably replaceable parts of a product are remanufactured. In either case, it is important to discern that in most contemporary applications of remanufacturing, ownership is transferred between remanufacturing cycles from the user back to the remanufacturer, and subsequently to another user that may or may not be the same as that from which the core was sourced. In this sense, remanufacturing creates entirely new use cycles—an important distinction from simple repair processes in which the original user retains ownership.

There are many processes that may be considered part of remanufacturing; the action itself can therefore be represented at a high level by a generic collection of seven sub-processes (Figure 2.4) that cores pass through in order to restore new-equivalent form and function [4].

However, the employment, order, and purpose of these different operations are not ubiquitous, but are rather dependent on individual product contexts. In practice, each remanufacturer may use a different group and sequence of remanufacturing steps depending upon a number of variable factors, including their core supply, customer demands, and industry-specific

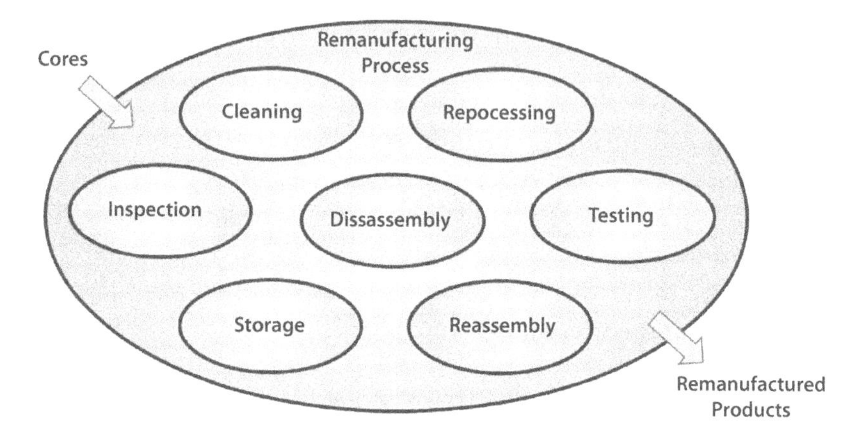

Figure 2.4 A generic remanufacturing process with seven general steps (adapted from [4]).

regulations. For example, cores with many subsystems may be disassembled first to allow component-level inspection, while simpler products may be inspected before disassembling their components. While remanufacturing is often described in literature with inspection occurring *after* cleaning and disassembly, this is not a prescribed mandate for remanufacturing. Indeed, there are cases in which inspection may reveal that cleaning and disassembly may not even be necessary, allowing remanufacturers to avoid the energy, material, labor, and opportunity costs of these steps—costs otherwise meaninglessly incurred at no benefit to the remanufacturer. This, however, has an equal and opposite case—many products may be impossible to accurately inspect if not cleaned. In all of this, it is clear that the remanufacturing process can be organized in many different ways according to context-specific requirements.

In general, remanufacturing is different in comparison with virgin (new) manufacturing in both process characteristics and operational logistics. For example, batch sizes that can be accommodated in remanufacturing are normally smaller than in conventional systems as a result of both the nature of reverse supply chains and the non-automated labor-intensity of remanufacturing processes that often require manual input [5]. However, as robot technologies improve to enable adaptable, unique, product-specific, and non-repetitive functions, there will be more opportunities to automate remanufacturing processes.

2.3.2 Who Remanufactures?

There are many products being remanufactured in many industrial sectors. In an investigation of the North American remanufacturing industry, Lund [6] created a database of more than 7,000 remanufacturing companies (6,600 in the United States and 491 in Canada) that had been verified as engaging in remanufacturing activity for at least two years. In an European Remanufacturing Network (ERN)[6] market study the European remanufacturing industry was estimated to involve 7,200 companies across Europe [7]. Naturally, these remanufacturing companies span may different industrial sectors, and cover myriad product types. Table 2.1 provides some examples, some of which are more frequently remanufactured than others.

Within the remanufacturing industry more broadly, there are three different types of agents that perform remanufacturing. These companies are usually divided according to their relationship to the entity that produces the virgin (new) product from which remanufactured products are made, i.e. the Original Equipment Manufacturer (OEM) [8]:

Original Equipment Remanufacturers: Companies that remanufacture their own products. These companies are sometimes called Original Equipment Remanufacturers (OERs), and could be a business unit of the OEM that performs the remanufacturing. In this case, the OEM/OER benefits from an established supply chain network, and can facilitate reverse logistics of products returning from workshops, retailer trade-ins, or end-of-lease contracts. OEMs/OERs also have the advantage of

[6] ERN - European Remanufacturing Network (ERN) - see www.remanufacturing.eu

Table 2.1 Examples of products currently being remanufactured, from A to Z.

Air conditioners	Excavation equipment	Office equipment	Tables
Air flow meters	Filling machines	Pallets	Toner cartridges
Alternators	Flat panel displays	Photocopiers	Truck engines
Bakery equipment	Forklift trucks	Pistons	Turbo charges
Brake calipers	Furniture	Power bearings	Undercarriage
Catalysts	Gaming equipment	Process fans	Valves
Chairs	Gear boxes	Pumps	Vending machines
Coffee machines	Heat exchangers	Quivers	Washing machines
Computers	Injectors	Robots	Wheel chairs
Compressors	Jet fans	Refrigerators	Wireless switches
Cubicles	Keyboards	Servo pumps	X-ray apparatus
Cylinders	Laptops	Smartphones	Yacht engines
Desks	Machine equipment	Soil compactors	Zimmer frames
Dishwashers	Machine tools	Start engines	
Diesel particle filters	Medical Equipment	Stapling machines	
Electrical motors	Notebooks	Steering pumps	

access to and ownership of design knowledge, spare parts, and service methodologies.

Contracted remanufacturers: Companies that are contracted to remanufacture products on behalf of other companies.

This means that the OEM normally owns the products and uses remanufactured products as a component of their business model, but does not perform the actual remanufacturing process. A contracted entity can then perform remanufacturing as a service for the OEM. For the contracted remanufacturer, this means that there is likely to be a fairly consistent stream of business with supply chain advantages, fewer working capital requirements (e.g. work in progress) and risks, and the company can expect to obtain assistance from the OEM in terms of replacement parts, design and testing specifications, and even tooling [8]. An example of this kind of contracted remanufacturer is UBD Cleantech, which remanufactures diesel particle filters in Höör, Sweden [9].

Independent remanufacturers: Companies that remanufacture products with little or no contact with the OEM, and must often purchase or collect cores from end users themselves. Independent remanufacturers often must also buy spare parts and develop product knowledge on their own. The typical independent remanufacturer is a private corporation with closely-held ownership [8]. Lund further suggests that this type of operation is an integrated one, in that it purchases cores, remanufactures them, and markets them under its own name (rather than leveraging the OEM brand) or for the private labels of others. In some cases, OEMs view independent remanufacturers as competition; even when the OEM does not itself remanufacture and is thus not threatened by direct competition, a fear of comparatively lower-priced remanufactured products cannibalizing sales of new (virgin) OEM products in some cases leads OEMs to deliberately advocate against remanufacturing through non-compatible design choices or inhibitory market mechanisms. Hammond *et al.* [10] posit that this relationship is in some cases tumultuous, but overall static, and therefore unlikely to grow towards cooperation in the future. However, independent remanufacturers are in some cases able to change their business model in effort to start collaborating with OEMs rather than compete with them as a means to facilitate the logistics of remanufacturing; in so doing, they effectively become a contracted remanufacturer. An example of an independent remanufacturer is Scandi-Toner, which remanufactures toner cartridges in Karlstad, Sweden [9].

2.3.3 Why Remanufacture?

There are many reasons companies remanufacture, such as profit, the environment, or policy compliance [11]. In a recent study from the European Remanufacturing Network (ERN), incentives for remanufacturing were summarized into seven points [7]:

- **Economics:** Remanufacturers see greater profit margins with service-based than "make and sell" businesses; product service systems promote remanufacturing rather than replacement.
- **Cost savings:** Remanufactured products usually cost less than new, attracting wide markets. While this differential varies both within and between sectors, a survey suggests the majority of remanufactured products sell for between 41- and 80-percent of the cost of a new product.
- **Access to cores:** Many remanufacturing businesses are supply-constrained, and increasing collection rates will enable remanufacturing to grow.
- **Reduced lead times:** Remanufactured products can have reduced lead times, minimizing disruption due to the failure of key systems.
- **Alternative businesses models:** These include rental and service-based offerings, which tend to lead to better relationships with customers and a more skilled and adaptable workforce. This can also lead to a reconfiguration of the supply chain to service the new business model.
- **Reduced risk of resource insecurity:** By keeping products "whole," remanufacturing reduces risks

associated with long supply chains and material criticality.
- **Environmental legislation:** The ELV Directive poses a problem for recycling, but could become an opportunity for remanufacturers—as of January 2015, 95-percent of a vehicle's material *must* be reused or recovered.

Specific answers amongst the 21 remanufacturing companies interviewed as a part of this ERN business model study varied, each with its own particularities. For example, economic benefits include the lower cost of remanufacturing compared to new manufacturing, which increase the potential for profit if the remanufactured product can be sold for a competitive price corresponding to its effectively equivalent value. However, customer expectations of discounting for non-virgin products usually mean that the prices are set lower. Of course, it is the market that decides the price of the remanufactured products; thus, price over time is not certain. However, most remanufacturers claim that products sell for prices 50-percent lower than new. While this counteracts the potential for higher profit margins, it creates customer benefit that, in turn, encourages further patronization of remanufacturers, creating business benefits as a result. Figure 2.5 illustrates examples of how 21 companies answered regarding the reduction in the price of remanufactured products [9].

The ERN study also included an income and cost comparison between manufacturing and remanufacturing. This revealed that in general, new product manufacturing achieves high revenues through selling a large volume of products, but incurs high costs for acquiring raw materials and making them into products. In contrast, revenues in remanufacturing come not from volume, but from the high extractable value of the product relative to the lower material and product manufacturing costs. However, work requirements in remanufacturing are more demanding because the unique process steps (e.g.

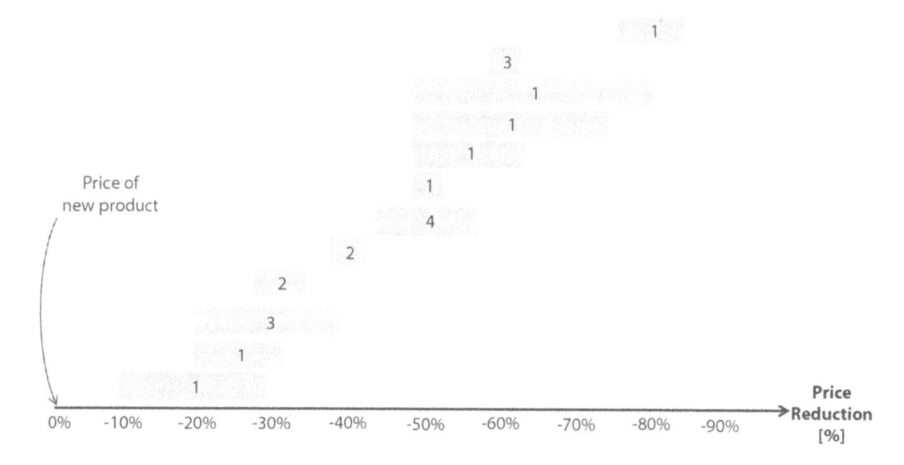

Figure 2.5 Reductions in prices of remanufactured products. The length of bars illustrates the price reduction range, while the number on the bars corresponds to how many remanufacturing companies have delivered the same answer [9].

disassembly, inspection, and diagnosis) are not well suited for automation. As a result, the number of required workers and skills flexibility is higher in remanufacturing, thus leading to greater personnel costs than in manufacturing. When balancing costs and revenues based on these tradeoffs, ERN suggests that profits can be equally large in both new manufacturing and remanufacturing. However, if the customer perception of remanufactured products increases to more accurately reflect their true value, remanufacturers could accordingly increase prices, thus increasing profitability significantly. This comparison is illustrated with an example in Table 2.2 [7].

In this example, volume-based revenue is clearly lower in remanufacturing. However, because the lifecycle value of remanufactured products in form, function, and durability is effectively equivalent to new products, it stands to reason that their market price should be competitive with new rather than steeply discounted. Such parity would increase potential remanufacturing revenue without increasing costs, thus improving profit margins significantly. Of course, this relies

Table 2.2 An example comparison of incomes and costs, manufacturer and a remanufacturer (adapted from [7]).

	Manufacturer	**Remanufacturer**
Turnover income	$ $ $ $ $ $ $	$ $ $ $ $
Raw material and energy costs	$ $ $ $	$
Personnel costs	$ $	$ $ $
Profit	$	$

on a shift in customer perceptions in which the full value of remanufactured products, including their environmental benefits over new products, is recognized with willingness to pay.

2.3.4 Why Not Remanufacture?

In order to remanufacture successfully, companies must overcome uncertainty and complexity challenges. Most of these are rooted in the uncertainties of the quantity and quality of returned products (cores). This uncertainty also creates variations in capacity requirements and process yield. Steinhilper and Butzer [12] explore the management of these uncertainties and find that their resultant impacts can include detrimental business effects such as longer lead times, inconsistent product quality, and thus a less profitable business. Beyond this fundamental challenge, ERN researchers suggest that the uptake of remanufacturing is inhibited by barriers that affect both remanufacturers and consumers differently. Amongst these, challenges affecting remanufacturers have especially significant impacts on growth. According to Parker *et al.* [7], these include:

- **Lack of technical information available to third-parties:** The knowledge necessary to remanufacture products effectively is not readily available to non-OEMs.
- **Legal ambiguity:** Lack of clarity over what remanufacturing entails. There is no clear guidance on the use of remanufactured components in new products, or whether remanufactured products must be declared "second-hand." Furthermore, there are issues regarding the effect on the legal and market status of remanufactured products from legislation such as the WEEE Directive, Waste Framework Directive, ELV Directive, Sales of Goods Act, REACH Regulation, RoHS Directive, and Energy Using Products Directive.
- **Definition of waste:** Ambiguity over whether the activities undertaken during remanufacturing are considered "waste processing" may affect remanufacturers. For example, the requirement to control and process products that are legally considered waste (which in many cases applies to product cores returned for remanufacturing) adds additional administrative and compliance costs to a business. There is a business risk where regulatory guidance is not provided.
- **Competition from lower-cost products:** This is widely cited as an issue across several remanufacturing sectors. The sale of anecdotally inferior-quality products whose virgin (new) status makes them attractive to customers undercuts the market for remanufactured products.
- **Lack of technically skilled engineers:** Skills shortages affect the industry as in manufacturing.
- **Poor design for remanufacturing:** Remanufacturing can sometimes be inhibited by poor design, particularly where remanufacturing is not embedded within the OEM culture.

- **Technology shifts:** As advances in materials and electronics occur, remanufacturers also need to make advances in their processing technologies to ensure that the end product matches the performance of new products. This includes energy efficiency, new materials, and the incorporation of more electric/electronic systems into traditionally mechanical-based products.
- **Reverse logistics costs:** The transport of large or bulky items can be a significant cost, which may prevent the remanufacture of certain goods or viability in sparsely-populated areas.
- **Cost and availability of storage space:** Storing large volumes of reused components is a large expense on remanufacturers.
- **Lack of remediation techniques:** In some sectors, technological advances in remediation are needed to ensure that remanufactured products match the performance of new products.

The majority of these barriers are not encountered in virgin (new) manufacturing, and certaintly not to the same extent if they are indeed shared. The fact that remanufacturing remains a viable business models in many markets despite these challenges suggests that alleviating them may in fact unlock immense potential for growth.

2.3.5 Why Buy Remanufactured Products?

The ERN study also sought to explore the benefits of remanufacturing to both customers and producers. Responses expectedly varied, but centered on common themes of business outcomes gained through customer benefits in price, quality, warranty, and uptime [9].

Most companies stated that remanufacturing is worthwhile compared to new manufacturing, as it enables them to offer a

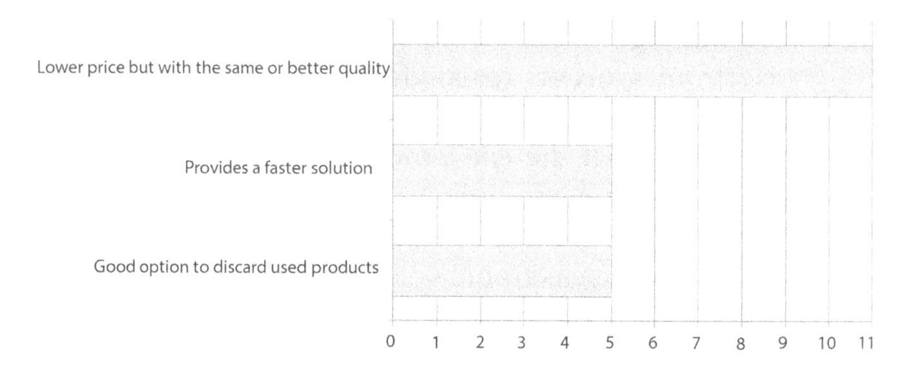

Figure 2.6 The most common answers regarding the remanufacturing companies' customer values [9].

lower price with equivalent quality. In one case, a remanufacturer related that their products provided even better quality than their mass-produced virgin counterparts. In another case, remanufactured products were the only option, since the original virgin alternative had gone out of production. Some companies stated that remanufactured products provide a faster solution than ordering new parts, meaning that up-time was higher and out-of-service times were lower. Remanufacturing was also seen as a good option to get rid of used products and thus avoid non-value-added inventory storage costs. Figure 2.6 illustrates these most common answers from the remanufacturing companies [9].

2.3.6 Why is Remanufacturing Good for the Environment?

Remanufacturing is often seen as an environmentally beneficial way of closing the flows of products, parts, and materials. One reason remanufacturing is attractive from an environmental point of view is that necessary processes of manufacturing new parts—e.g. material extraction, material manufacturing, part manufacturing, and product assembly—are fully or partly avoided in remanufacturing by salvaging the form and function of an already finished product at EoU. As a result, the impacts of these processes in material depletion, energy consumption, emissions generation, and waste creation are all mitigated significantly, if not completely avoided.

Figure 2.7 illustrates how a product's environmental impact accumulates during its lifecycle. For energy-using products, most of the environmental impact typically occurs during use. Conversely, products that do not use energy during use create most of their environmental impact during manufacturing and at the EoL. In either case, remanufacturing extends the physical life of a product for a complete additional useful life after each cycle, spreading the lifecycle environmental impacts of manufacturing a single product over a greater period of time, and thereby reducing impacts markedly relative to traditional linear use models.

The figure also illustrates that the ability of outside actors to influence a product's lifecycle environmental impacts decreases significantly as the lifecycle progresses. In this, the greatest potential to influence (i.e. reduce) the impacts created in a product's use and EoL stages is early in the design phase. This means that early efforts in designing a product to improve its

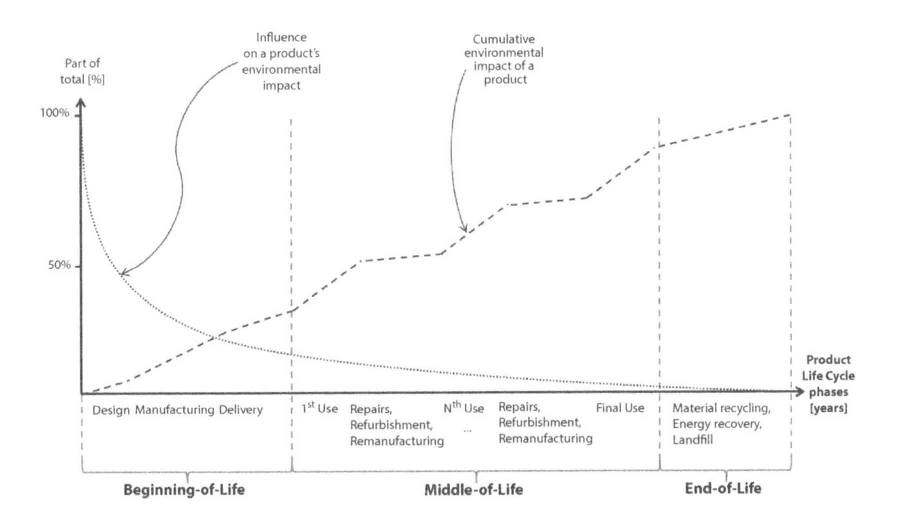

Figure 2.7 The accumulated environmental impact during a product's life. In addition, there is also a graph showing where the most influence on the environmental impact occurs (adapted from [13]).

environmental performance during use, as well as in designing a product to be remanufactured at EoU, can have exponential returns on environmental impacts [14].

In a literature review by Sundin and Lee [15], 17 surveys were gathered showing that, from an environmental perspective, remanufacturing preferable to both new manufacturing and material recycling. While material and energy resources savings over new manufacturing are clear, a comparison to material recycling is somewhat more complex. Material recycling is, of course, preferential to landfill disposal, but its environmental impacts and relative value are less preferable than remanufacturing. In salvaging only the constituent material of a product, recycling fails to preserve the value of product form and function, and thus requires subsequent manufacturing processes to turn that secondary raw material back into a working product. Savings in the material, energy, labor, and emissions intensity of manufacturing are therefore not achieved with material recycling—rather, only the impacts of raw material extraction and refining are avoided. As a result, a summary of environmental impact surveys suggests that compared to *both* virgin (new) manufacturing and material recycling, remanufacturing:

1. **Uses fewer material resources**: less material input is required to remanufacture than to manufacture new products, regardless of whether new primary or secondary material is used.
2. **Has a lower greenhouse effect (less CO_2 emissions)**: avoiding the processes in material extraction, component manufacturing, *and* product assembly reduces the amount of energy consumed, and therefore the volume emissions produced.
3. **Facilitates safer handling of hazardous materials**: bringing back products for remanufacturing also brings back toxic and hazardous material that otherwise could end up in forests, landfills or recycling equipment and recycled materials.

Remanufacturing thus creates significant environmental benefits in key areas by both avoiding the direct CO_2 emissions of primary manufacturing and diverting large number of products materials away from landfill (remanufactured products typically contain up to 80-percent of their original material content). The latter of these then propagates further environmental benefits by reducing landfill emissions as well as the pace and effects of land use change. A study of European remanufacturing impacts (Table 2.3) suggest that benefits in these area include 8.3 million metric tons of CO_2 savings and 2.3 million metric tons of avoided landfill material [7].

Table 2.3 Material and CO_2 savings from industry in Europe from remanufacturing [7].

Industry sector	Materials (k tonnes (metric tons))	CO_2 equivalents (k tonnes (metric tons))
Aerospace	136	356
Automotive	902	3,298
EEE	150	177
Profit	1	1
Furniture	76	131
Heavy duty and off-road equipment	855	3,458
Machinery	35	393
Marine	15	40
Medical equipment	22	58
Rail	69	344
Total	2,260	8,255

As a key example, the Automotive Parts Remanufacturers Association (APRA) Europe branch[7] estimates that remanufacturing automotive components yields some 88-percent material savings and 56-percent lower energy usage compared to manufacturing a new product, with an associated 53-percent decrease in CO_2 emissions.

One must also remember that in order to remanufacture a product, it must be manufactured in the first place. Further, new product technology is rapidly advancing, often improving the use-phase environmental performance of new products at a likewise impressive pace. In this, it is imperative to recognize that primary manufacturing and remanufacturing must always coexist, and that their environmental impacts will likely continue to sway in a dynamic balance. In order to minimize overall impacts of the industrial economy, however, we must strive to achieve two key factors in this balance. First, we must optimize product designs for upgradability to allow integration of advancing technology on older products, so that they may remain useful and competitive (and thus offset the need for primary manufacturing) for as long as possible. However, we must also strengthen strategies for remanufacturing decision-making to identify where deference to a new product with vastly improved use-phase performance may indeed lower lifecycle impacts beyond the reach of remanufacturing. Parameters like the fuel efficiency, for example, can in this regard alter lifecycle environmental impacts significantly in new products in ways that remanufacturing legacy components simply cannot equal.

2.4 Statements from Industry and Conclusions

2.4.1 Statements from Industry

Within the automobile industry remanufacturing is a natural activity while for other industries, for example the smart phone industry, remanufacturing is still in its infancy. In either

[7] http://www.apra-europe.org/dateien/News/News2015/APRA_Position_Paper.pdfC.

case, the remanufacturing industry has ideas on how to make our current linear material flows move toward a more circular economy. Below are two statements from practitioners within the remanufacturing industry that exemplify this potential.

The first is from APRA[8], which states the following about the future:

> *"As people grow more and more concerned about the environment, we must seek public policies that will encourage even more remanufacturing. There are enough social and environmental benefits to justify remanufacturing. Imagine the added benefits to society if* ***everything*** *we buy could be remanufactured, from small appliances to lawn mowers. Imagine if products were originally manufactured with the sole purpose of being rebuilt and not thrown away."*

The second is from an environmental and quality manager of IT equipment remanufacturer Inrego[9], who states the following about remanufacturing operations and the importance of customer perceptions:

> *"In essence, apart from the obvious logistic function of moving IT equipment from the first user to the next, Inrego reinjects trust into the product by sorting, testing, repairing, and renewing the warranty. That said, the role as communicator of why remanufactured products are just as good as new also becomes important"*- Erik Pettersson, Inrego AB.

2.4.2 Remanufacturing as the Heart and Lungs of the Circular Economy

This chapter has described both the circular economy and remanufacturing, along with their ideas and goals. In their descriptions of the circular economy, different authors stress

[8] http://www.apra-europe.org/dateien/News/News2015/APRA_Position_Paper.pdf.

[9] The company Inrego AB is described in Sundin et al. [9] and at www.inrego.se.

a common theme: that circular loops should be kept as tight as possible around the user in order to achieve economic and ecological benefits, minimizing the material left for outer loops, incineration, or landfill. In order to achieve this, strategies beyond simple reuse and material recycling are imperative to preserve the value of product function, which is rapidly becoming the central focus of customer demands. In this light, remanufacturing is essential for the circular economy.

However, there are still some hurdles to overcome in order to increase global uptake of remanufacturing in today's globally industrialized society. The most challenging of these—and therefore those that deserve the greatest portion of our attention—are customer acceptance, legal issues, company policies, and an understanding of what the benefits of remanufacturing really are. Despite these remaining challenges, many companies in different industry sectors manage to create economically and ecologically sustainable businesses from remanufacturing operations. Alleviating these barriers, then, may unlock an unprecedented growth in remanufacturing that can propel us into the future of a circular economy. While possible, this requires a fundamental paradigm shift in the mindset of both manufacturing companies and their customers that is not only aware of, but truly values the triple-tired benefits of the circular economy. By pursuing this paradigm shift, the adoption of circular principles will surely accelerate, and naturally with it the uptake of remanufacturing as a means to achieve it.

The role of remanufacturing in the circular economy is thus clear. As Steinhilper [5] states, remanufacturing is the ultimate form of recycling. But beyond this, as APRA [16] posits, remanufacturing is simply the best practice of a circular economy. And while the European Remanufacturing Council (ERC) [17] contend that remanufacturing is the *backbone* of circular economy, I would suggest instead that remanufacturing is the *heart and lungs* of circular economy. In true circular fashion, where the heart pumps physical products back out on the market, fills them with added value at the lungs by the reprocessing step, and send them further to provide the benefit of life to the system's dependents—our global society—after which time they return to be used again and again in a circular economy.

References

1. EU, The EU report "Towards a circular economy: A zero waste programme for Europe", http://eur-lex.europa.eu/legal-content/EN/TXT/?uri = CELEX:52014DC0398, 2014.
2. Sundin, E. and Bras, B., Making Functional Sales Environmentally and Economically Beneficial through Product Remanufacturing. *J. Cleaner Prod.*, 13, 9, 913–925, 2005.
3. Sundin, E., Circular economy and design for remanufacturing, Chapter 18, in: *Designing for the Circular Economy*, M. Charter (Ed.), pp. 186–199, Routledge, Abingdon, UK, ISBN: 978-1-138-08101-7, 2019.
4. Sundin, E., *Product and Process Design for Successful Remanufacturing*, Linköping Studies in Science and Technology, Dissertation No. 906, Department of Mechanical Engineering, Linköping University, SE-581 83, Linköping, Sweden, 2004.
5. Steinhilper, R., *Remanufacturing - The Ultimate Form of Recycling*, Fraunhofer IRB Verlag, Stuttgart, ISBN: 3-8167-5216-0, 1998.
6. Lund, R., The Remanufacturing Database, August 17, 2012, accessed from: http://www.bu.edu/reman/The%20Remanufacturing%20Database.pdf, 2012.
7. Parker, D., Riley, K., Robinson, S., Symington, H., Tewson, J., Jansson, K., Ramkumar, S., Peck, D., Remanufacturing Market Study, report from the Horizon 2020 project: ERN – European Remanufacturing Network, Grant Agreement No. 645984, accessible from www.remanufacturing.eu, 2015.
8. Lund, R., *Remanufacturing: The Experience of the United States and Implications for Developing Countries*. CPA/83-17, The World Bank, Washington, D.C., 1983.
9. Sundin, E., Sakao, T., Lindahl, M., Kao, C.C., Joungerious, B., Ijomah, W., Map of Remanufacturing Business Model Landscape, report from the Horizon 2020 project: ERN – European Remanufacturing Network, Grant Agreement No. 645984, accessible from www.remanufacturing.eu, 2016.
10. Hammond, R., Amezquita, T., Bras, B., Issues in Automotive Parts Remanufacturing Industry: Discussion of Results from

Surveys Performed Among Remanufacturers. *J. Eng. Des. Autom.*, 4, 1, 27–46, 1998.
11. Östlin, J., Sundin, E., Björkman, M., Importance of Closed-Loop Supply Chain Relationships for Product Remanufacturing. *Int. J. Prod. Econ.*, 115, 2, 336–348, 2008.
12. Steinhilper, R. and Butzer, S., Remanufacturing, Closed-Loop Systems and Reverse Logistics, in *Remanufacturing*, Nasr, N (ed.), Wiley-Scrivener, Beverly, 2020.
13. Lindahl, M., Sundin, E., Product Design Considerations for Improved Integrated Product/Service Offerings, in: *Chapter 37 in Handbook of Sustainable Engineering*, J. Kauffman and K.M. Lee (Eds.), pp. 669–689, Springer, Dordrecht, ISBN: 978-1-4020-8938-1, 2013.
14. Kerr, W. and Ryan, C., Eco-efficiency gains from remanufacturing - A case study of photocopier remanufacturing at Fuji Xerox Australia. *J. Cleaner Prod.*, 9, 75–81, 2001.
15. Sundin, E. and Lee, H.M., In what way is remanufacturing good for the environment? Chapter in *Design for Innovative Value Towards a Sustainable Society, Proceedings of EcoDesign 2011: 7th International Symposium on Environmentally Conscious Design and Inverse Manufacturing*, Kyoto, Japan, November 30–December 2, pp. 551–556, Springer, Dordrecht, ISBN 978-94-007-3010-6, 2011.
16. APRA, Remanufacturing as best practice of the Circular Economy, a position paper of the Automotive Parts Remanufacturer Association - Europe, accessible from: https://apra.org/page/RemanResources#soc, 2019.
17. ERC, European Remanufacturing Council, Information on website: https://www.remancouncil.eu/, 2019.

Further Reading

The ERN project website: www.remanufacturing.eu.
Environmental studies gathered by APRA: www.apra-europe.org/main.php?target=environment.
Östlin, J., Sundin, E., Björkman, M., Product Life-Cycle Implications for Remanufacturing Strategies. *J. Cleaner Prod.*, 17, 11, 999–1009, 2009.

3

Remanufacturing Business Models

Gilvan C. Souza

Kelley School of Business, Indiana University, Bloomington, IN

Abstract

In this chapter I present some key business considerations for a firm that performs remanufacturing. In particular, I discuss two key decisions: (1) whether to introduce a remanufactured product, and its competitive implications; and (ii) used product (core) acquisition. In the first decision, I discuss key aspects to be analyzed before an original equipment manufacturer (OEM) decides to introduce a remanufactured version of its product: remanufacturing technology and underlying costs, market acceptance of remanufactured products, used product acquisition logistics, environmental considerations, brand protection, competitive landscape, cannibalization, and market expansion considerations (a function of price points of new and remanufactured products). I then present a simple optimization model that sheds light onto the decision, showing that the OEM's optimal decision to remanufacture can be characterized by a simple rule of thumb, involving unit remanufacturing and product acquisition cost relative to the unit cost for producing a new product, and consumers' discount factor for the remanufactured product relative to the corresponding new one. I then extend the model to the case where remanufacturing is performed by third party remanufacturers. I then discuss simple frameworks and decision models for core acquisition, which, due to variable quality condition of cores, involve a key trade-off: a large core acquisition quantity increases direct

Email: gsouza@indiana.edu

Nabil Nasr (ed.) Remanufacturing in the Circular Economy (61–84)

acquisition costs, but decreases unit remanufacturing cost, as the firm can be more selective in deciding which cores to remanufacture.

Keywords: Remanufacturing, circular economy, closed-loop supply chains, reverse logistics, sustainable operations

3.1 Introduction

A firm engaging in remanufacturing needs to make several decisions, at different levels: strategic, tactical, and operational, depending on the time horizon of their impact in future operations. An example of strategic decision is network design: size and location of remanufacturing facilities, collection centers, and distribution centers, among others. For an original equipment manufacturer (OEM), an important strategic decision is whether it should offer a remanufactured version of its product in the first place, as opposed to recycling used products for materials recovery. When an OEM offers a remanufactured product at a discount relative to the new product price, it captures additional customers who are not willing to pay the full price for the new product, however, it can also cannibalize on sales of the new product. I discuss this decision in this chapter; in particular I also discuss pricing of remanufactured products, which determines the magnitude of the market expansion and cannibalization effects. I also offer some perspectives on third-party remanufacturing. A key tactical decision in remanufacturing is used product acquisition: how many used products, or cores, to acquire at a given period, at which price, and at which quality levels. Operational decisions, such as job shop scheduling, are made daily, and have only short-term impact on future operations. I focus on strategic and tactical decisions in this chapter, as they drive most of the profits and environmental impacts. The reader interested in operational issues is referred to [12].

To illustrate, consider Cummins, the original equipment manufacturer (OEM) of diesel engines based in Columbus, Indiana. Cummins remanufactures end-of-use engines or engine parts, which have undergone a full usage cycle with a customer, and must be remanufactured in order to be functional again. For a diesel engine or part, the remanufacturing process is quite extensive, and consists of five different steps: full disassembly, thorough cleaning of each part (through a combination of chemical cleaning and sand blasting), reconditioning to restore a part's functionality to that of a new part, reassembly (including the addition of new parts to replace parts that cannot be remanufactured), and testing. Remanufactured parts (or engines) typically sell at a 35% average discount relative to the corresponding new part. Cummins has a trade-in program, which is common in automotive remanufacturing, where customers buying new or remanufactured parts receive discounts if they return their core. Such a program feeds the remanufacturing process, and allows Cummins to restrict access to cores by third-party remanufacturers (3PR). In the case of Cummins, the core must be remanufactured to be functional again; other types of remanufacturing, such as remanufacturing of one-generation old cell phones, or refurbishing of consumer returns, the remanufacturing process can be considerably simpler, and may not even involve full disassembly. Remanufacturing of one-generation old cell phones, for example, can be as simple as performing cosmetic repairs, full testing, software updates, and battery replacement.

3.2 Should an OEM Remanufacture?

Should an original equipment manufacturer (OEM) offer a remanufactured version of its product? There are a variety of factors that are important when making such a decision, such as remanufacturing technology, remanufacturing cost, presence of 3PRs, market acceptance of remanufactured products, environmental implications, alternatives to remanufacturing (for cores), impact on brand, ease of core collection (access to cores, and transportation costs to the remanufacturing facility), possible cannibalization of sales of new products, and many others. I categorize the discussion into factors that favor remanufacturing, and those that may not favor remanufacturing, and then present a simple mathematical model that sheds light in this decision, and that considers many of these factors.

Remanufacturing technology and remanufacturing cost are two closely linked factors. Engine remanufacturing, for example, is a high-tech operation, as maintaining the tight tolerances in the cylinders, pistons, and cylinder head given variability in core condition requires sophisticated technology, with a high level of automation. Despite this technical challenge, the unit (variable) cost of remanufacturing is almost always lower than the unit (variable) cost of manufacturing a new product from raw materials, due to the significant savings in energy and materials, especially if the core acquisition cost is reasonable.

The savings in energy and materials typically make the environmental case for remanufacturing in a circular economy (I discuss later examples where that is not the case.) From the industrial ecology literature, the standard tool to measure the total environmental impact of a product in its entire life cycle, including raw materials sourcing, manufacturing, distribution, use with customers, and end-of-life, is life-cycle assessment (LCA). An LCA study of diesel engines concluded that the remanufacturing process consumes between 68% and 83% less energy, and between 26% and 90% less materials than a regular manufacturing process [14]. If the remanufactured engine has similar fuel consumption as the corresponding new engine, which occurs if the remanufacturing process upgrades the technology of the old engine to that of a new engine, then the environmental benefits of remanufacturing, from a life cycle perspective, are clear.

When an OEM remanufactures (or subcontracts remanufacturing to a trusted third-party; I don't make this distinction in the economic models that follow), it sells a remanufactured version of its product that is certified by the OEM. This can help protect the OEM's brand, as it is difficult for a customer to identify if a quality problem in a product remanufactured by a 3PR originates from the 3PR's remanufacturing process, or from the original OEM product design. In fact, by offering a remanufactured version of its product, the OEM expands its customer base by reaching a segment of customers who would not be willing to pay the full price for the new product offered by the OEM, and would instead buy a cheaper competitive product, or a remanufactured product offered by a 3PR. Thus, remanufacturing helps an OEM fight competition both from 3PRs, as well as from lower-end substitutes.

Not all consumers at all markets accept remanufactured products as substitutes to new products. Hauser and Lund [9] argue that industrial customers are more sophisticated than consumers in their knowledge of remanufacturing processes, and thus have a higher acceptance of the remanufactured product as an alternative to a new product. Indeed, one of the largest OEM remanufacturers is Caterpillar, as it sells primarily to businesses. For consumers, however, remanufacturing can be a "black box," and this uncertainty about the quality of the remanufacturing process may translate into lower acceptance of remanufactured products, as demonstrated by several recent empirical studies [1, 17].

Product acquisition is a significant issue for remanufacturers in several industries. In fact, in automotive remanufacturing, OEMs sometimes need to practice "seeding", where it sells a new product as a remanufactured product, due to a lack of quality cores for remanufacturing. Customers expect the availability and lower price of the remanufactured product, and the inherent uncertainty in core acquisition in these industries may require seeding to keep the remanufacturing process operating. In addition, in cases where remanufacturing is labor intensive, competitive forces may lead OEMs to locate remanufacturing facilities in areas with lower labor cost; if that is the case, then transportation costs (for cores) may add up, particularly for bulky products. Network design optimization models can help in this location decision for the remanufacturing facility, and the reader is referred to [15] for an example of such decision model.

From an environmental perspective, there are also cases that do not favor remanufacturing when one considers the entire life cycle. Some LCA studies (e.g., [13]) have shown that refrigerators, for example, consume in excess of 80% of their life cycle energy in the stage of use with consumers. Despite savings in materials and energy from remanufacturing, the LCA finding suggests that, from an environmental standpoint, old refrigerators should be recycled for materials recovery instead of being remanufactured, due to higher energy efficiency of more modern refrigerators. In other words, due to its older technology, a remanufactured refrigerator consumes more energy during its life cycle than a newly manufactured refrigerator with higher energy efficiency.

Finally, the addition of a remanufactured product to the OEM's product line has two strategic implications. First, there is the market expansion effect discussed before, as the lower-priced remanufactured product reaches a new customer segment that is not willing to pay the higher price of the OEM's new product. Second, there is also a cannibalization effect, as some consumers who would have previously purchased the new product switch to buying the remanufactured product. Thus, pricing of the two products is a critical strategic decision by the OEM, and I now offer an analytic model for such a decision, and some insights.

3.2.1 A Model to Answer the Question

The basic modeling framework to answer this question uses a vertical differentiation model, where the two products, remanufactured and new, are differentiated in terms of their perceived quality by customers. To keep it simple, I consider the simplest possible case of a monopolist OEM, which manufactures a (new) product and considers the introduction of a remanufactured version of the same product. This is a single period problem, which can be thought of as a period in a long time horizon, where a period could be a typical useful life of the product in consideration. This scenario, and its variations, has been extensively considered in the literature; see for example [3] and [5]. Competition can be included, as discussed below, but it significantly complicates the model, and additional insights are limited. Unit variable production cost for the new product is c_n, and unit variable cost for the remanufactured product is $c_r < c_n$. The price of the new product is p_n and the price of the remanufactured product is $p_r < p_n$, which are decision variables. The first assumption in the model is:

Assumption 1: Consumer willingness-to-pay (WTP) W for the new product is uniformly distributed in the interval $[0, B]$.

A consumer's WTP W is the maximum price he or she is willing to pay for the new product. Assuming that W is uniformly distributed in $[0, B]$ means that the probability that a consumer has a WTP W higher than a, for example, is $\frac{B-a}{B}$. If there is no remanufactured product, and the OEM's price for the new product is p_n, then only consumers with WTP higher than p_n buy the product. This means that if the total market size is M, then the quantity q_n sold by the OEM is equal to $q_n = M \cdot \Pr\{W \geq p_n\} = M\frac{B-p_n}{B} = M \cdot (1 - p_n/B)$. This is the familiar linear demand curve, illustrated in Figure 3.1(a), which is a result of the uniform distribution in Assumption 1.

An inverted S-shaped (non-linear) demand curve is perhaps more realistic than the linear demand curve: As prices are very low, almost all consumers buy the product, and thus demand decreases very slowly as price increases from very low levels. At a certain price point, however, prices become significant enough that less and less consumers buy the product as prices increases (i.e., there is a steeper slope of the demand curve). At very high prices, only very loyal customers buy the product, and thus demand decreases slowly as price increases. A logit demand curve simulates this behavior, and it can be obtained if

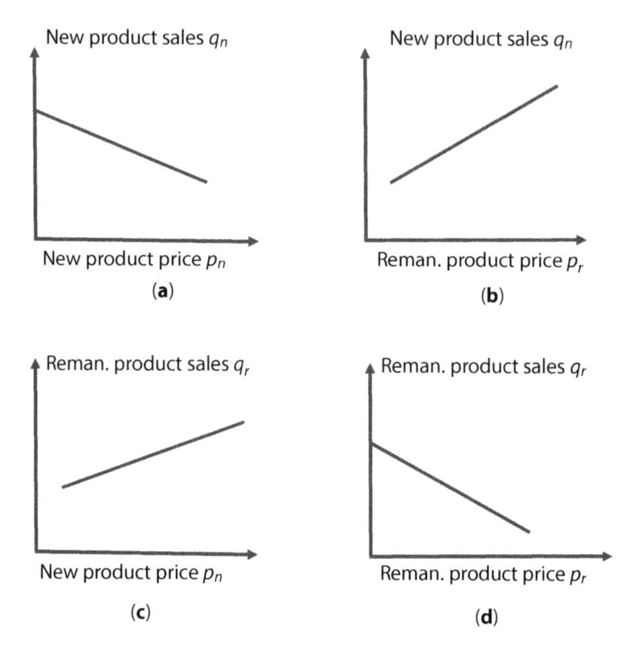

Figure 3.1 New product demand as a function of (a) new product price, and (b) remanufactured product price. Remanufactured product demand as a function of (c) new product price, and (d) remanufactured product price.

one assumes a Gumbel WTP distribution (instead of uniform, as in Assumption 1): $\Pr\{W \leq x\} = (1 + e^{-x})^{-1}$. The logit demand curve, however, is not analytically tractable. Linear demand curves are an approximation of the logit demand curve if prices are allowed to vary within narrower limits, that is, prices vary in the middle portion of the logit demand curve.

Assumption 2: A consumer of type W (i.e., WTP for new product equal to W) has a WTP for the remanufactured product equal to δW, where $0 \leq \delta \leq 1$.

Assumption 2 indicates that remanufactured and new products are vertically differentiated—consumers prefer a new unit to a remanufactured unit for the same price, and δ is a product-specific parameter, constant across consumers. (This is in contrast to horizontally differentiated products, such as green vs. blue shirts, where some consumers prefer green whereas other consumers prefer blue shirts). If $\delta = 0$, consumers do not consider the remanufactured product as an alternative to the new product. If $\delta = 1$, consumers view the new and remanufactured units as perfect substitutes, and are indifferent between new and remanufactured products at the same price point. Clearly, most products are in between these two extremes, so that $0 < \delta < 1$; in fact there is considerable empirical evidence that consumers *perceive* remanufactured products to be of lower quality than new products. Hauser and Lund [9] report 35-55% *average* discounts for a remanufactured product relative to new. Guide and Li [7] use online auctions to derive approximate values $\delta \sim 0.85$ for power tools, and $\delta \sim 0.90$ for Internet routers. Subramanian and Subramanyam [16] compare online prices of new and remanufactured products, and find that average prices of remanufactured products relative to their corresponding new counterparts (a proxy for δ) range from 0.60 (for video game consoles) to 0.85 (for some consumer electronics).

Assumption 2 allows us to derive the demand curves when there are new and remanufactured products as follows. A consumer of type W has a net utility $U_n = W - p_n$ for the new product and $U_r = \delta W - p_r$ for the remanufactured product. If $U_n > U_r$, then the consumer derives a higher net utility from the new product than for the remanufactured product, and the consumer buys the new product. If one solves this simple linear inequality for W one finds that all consumers with $W \geq \frac{p_n - p_r}{1-\delta}$ buy the new product. Conversely, if $U_n < U_r$ and $U_r > 0$, the consumer buys a remanufactured product. Solving the two linear inequalities for W implies that all consumers with W satisfying $\frac{p_r}{\delta} < W < \frac{p_n - p_r}{1-\delta}$ buy the remanufactured product. Finally, all consumers with negative net utility for the remanufactured product do not buy anything, which means consumers with WTP $W < \frac{p_r}{\delta}$ do not buy anything. This market segmentation is shown in Figure 3.2.

To derive the demand functions from the market segmentation in Figure 3.2, we use Assumption 1, which

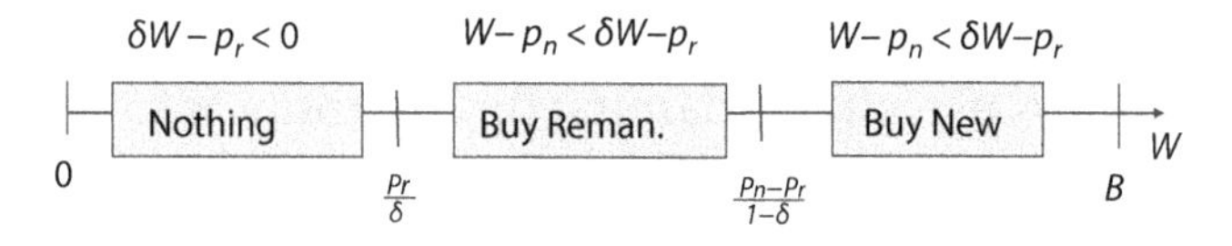

Figure 3.2 Market segmentation with new and remanufactured products.

states a uniform distribution for W. It then follows that the quantity of new products sold, given prices pn and pr is $q_n = M \cdot \Pr\left\{W \geq \frac{p_n - p_r}{1-\delta}\right\} = M \cdot \left(B - \frac{p_n - p_r}{1-\delta}\right)/B$. By the same logic, the quantity of remanufactured products sold is then $q_r = M \cdot \Pr\left\{\frac{p_r}{\delta} < W < \frac{p_n - p_r}{1-\delta}\right\} = M \cdot \left(\frac{p_n - p_r}{1-\delta} - \frac{p_r}{\delta}\right)/B$. Thus, the demand functions when there are new and remanufactured products are:

$$q_n(p_n, p_r) = M \cdot \left(1 - \frac{p_n}{B(1-\delta)} + \frac{p_r}{B(1-\delta)}\right) \tag{3.1}$$

$$q_r(p_n, p_r) = \frac{M}{B(1-\delta)}\left(p_n - \frac{p_r}{\delta}\right) \tag{3.2}$$

These demand curves are illustrated in Figure 3.1. Figure 3.1(a) indicates that demand for the new product decreases linearly in the new product's price (for a given remanufactured product price), and Figure 3.1(b) shows that demand for the new product increases linearly in the remanufactured product's price (for a given new product price). Similarly, Figure 3.1(c) indicates that demand for the remanufactured product increases linearly in the new product's price (for a given remanufactured product price), and Figure 3.1(d) shows that demand for the remanufactured product decreases linearly in the remanufactured product's price (for a given new product price). That is, new and remanufactured products are (imperfect) substitutes.

The OEM thus solves the following optimization problem, where the decision variables are the prices of new and remanufactured products, which determine the respective demands:

$$\max_{p_n, p_r} \text{Profit} = (p_n - c_n) \cdot q_n(p_n, p_r) + (p_r - c_r) \cdot q_r(p_n, p_r) \tag{3.3}$$

$$\text{subject to } 0 \leq q_n(p_n, p_r) \leq q_r(p_n, p_r) \tag{3.4}$$

The optimization problem (3.3)-(3.4) captures several aspects of the OEM's decision, discussed previously. It captures the market's acceptance of remanufactured products through the parameter δ, the relative costs of remanufacturing and remanufacturing technology through the parameters c_r and c_n, and the degree of market expansion and cannibalization through the demand curves (3.1)-(3.2). Product acquisition costs are embedded in c_r. It does not consider 3PR competition, and I present later the case where the OEM competes with a 3PR.

Note that in (3.3) one must substitute the demand curve expressions (3.1)-(3.2), which results in a quadratic objective function with two decision variables (p_n and p_r). The constraint (3.4) simply indicates that the OEM has to sell more new products than remanufactured products, to have a sustainable remanufacturing program, and it is a linear inequality in the decision variables p_n and p_r, considering (3.1)-(3.2).

Although the problem appears to be simply about setting prices, its structure also provides the answer to our question of whether the OEM should remanufacture. This is because prices imply quantities sold, and prices can be set in a way that results in no demand for remanufactured products. Indeed, it can be shown (see, e.g., [3]) that the optimal solution to this problem has a surprisingly simple structure, which is shown in Figure 3.3. Figure 3.3 indicates that the OEM remanufactures—that is, it sets the optimal prices in such a way that it results in non-zero demand for remanufactured products—if $c_r < c_n\delta$. The optimal prices themselves are given by $p_n = (B + c_n)/2$ and $p_r = (\delta B + c_r)/2$.

This result is simple but quite insightful. If the unit cost of remanufacturing c_r, which includes product acquisition cost, is low enough relative to the unit production cost for the new product c_n, then remanufacturing is attractive to the OEM, and it should offer a remanufactured version of its product. Notice that here "low enough" is related to the market's acceptance for remanufactured products δ, and so this model shows that remanufacturing for an OEM is not simply a question of costs, but it depends highly on the market's acceptance of remanufactured products, as it determines the market expansion and cannibalization effects.

Moreover, prices are set in a way that the market expansion effect outweighs the cannibalization effect. Remanufacturing technology, as well as a careful design of the reverse logistics

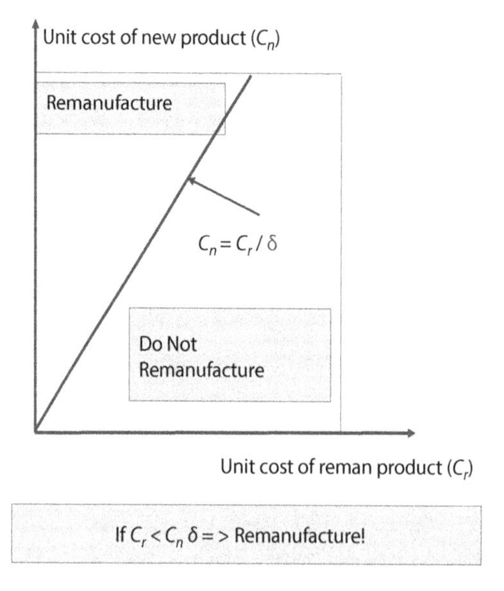

Figure 3.3 Solution to the OEM remanufacturing problem.

network for core acquisition, can lower the unit remanufacturing cost c_r and make it attractive for an OEM to offer a remanufactured version of its product. Conversely, the OEM can educate consumers on the thorough nature of remanufacturing processes, or offer similar warranties as new products, thus improving consumers' quality perception of remanufactured products—a higher δ—again, making remanufacturing and the circular economy more attractive.

3.2.2 3PR Competition

If the OEM only sells the new product, and a 3PR offers the remanufactured product, then the market segmentation of Figure 3.2 remains unchanged. Now, the OEM selects the new pro duct price p_n, whereas the 3PR chooses the remanufactured product price p_r. The difference with respect to (3.3)-(3.4) is that now we have a game-theoretical model where the OEM maximizes the first term and the 3PR maximizes the second term in equation (3.3) by setting their respective optimal prices p_n and p_r, with constraint (3.4) being added to the 3PR's problem. It can be shown (see, e.g., [3]) that the 3PR competes with the OEM—i.e., offers its remanufactured product—even for a higher unit remanufacturing cost than $c_n\delta$, which is the threshold from Figure 3.3 for OEM remanufacturing. In other words, Figure 3.3 indicates that the OEM does not remanufacture if $c_r > c_n\delta$. A 3PR, however, may still remanufacture if $c_r > c_n\delta$, within reason. (Technically speaking, the 3PR remanufactures as long as $c_r < \frac{\delta c_n + \delta(1-\delta)B}{2-\delta}$, and this threshold is larger than δc_n. Thus, remanufacturing is attractive for the 3PR in a wider range of c_r than for the OEM.)

Thus, remanufacturing is in general more attractive for the 3PR than for the OEM (without the threat of 3PR entry). The reason is simple: when a 3PR remanufactures, it only has a "market expansion" effect—there is no cannibalization effect, and thus the 3PR enters as long as it can profitably sell to even a small segment of customers. When the OEM remanufactures, it has to balance the market expansion effect with the cannibalization effect, and so it prefers to remanufacture (in the absence of a credible threat of entry by a 3PR) only when the market expansion effect outweighs the cannibalization effect.

The discussion above considers only the case where there are only two products: a new product offered by the OEM, and a remanufactured product offered by either the OEM, or by the 3PR, but not both. If there are two remanufactured products: one offered by the 3PR, and one offered by the OEM, then we need demand curves that consider three substitutable products (OEM new, OEM remanufactured, 3PR) instead of two. This would require an assumption about how consumers perceive remanufacturing by a 3PR compared with remanufacturing by the OEM. Most likely, consumers would prefer the remanufactured product by the OEM to the remanufactured product by the 3PR, everything else equal (prices, warranties, lead times, service). As a result, one would then need to consider an additional parameter: the discount factor that consumers have for the remanufactured product by the 3PR compared to the remanufactured product by the OEM (similarly to δ, between new and remanufactured products). Such game-theoretical model is considerably more difficult to analyze for finding optimal prices. Besides the additional assumption for building the demand functions indicated above, one would also need to specify how cores are split between the OEM and the 3PR, which can be a significant issue in certain industries, such as automotive, where cores are not abundant. Despite these difficulties, one can still show that, when there is a credible threat of 3PR entry, the OEM remanufactures even if $c_r > c_n\delta$, in order to deter entry by the 3PR and/or alleviate its competition.

3.2.3 Other Strategic Considerations

Trade-ins, leasing, and relicensing fees: The economic models discussed in the previous section do not explicitly specify the manner by which the firm obtains its cores for remanufacturing. They simply consider that remanufacturing is constrained by the amount of new products previously sold by an OEM. The OEM can employ different strategies for obtaining its cores. As in the Cummins case discussed previously, the OEM can offer a trade-in program, whereby it provides a discount in the price of a product it sells if the customer returns his/her core. Leasing is also an attractive mechanism for obtaining cores for remanufacturing, as well as controlling the secondary market. Agrawal *et al.* [2] compare leasing relative to selling as strategies for the OEM, and they use assumptions 1 and 2 to model consumer preferences between new and used products. With leasing, used products are remanufactured and resold by the OEM, whereas in the case of selling, consumers trade used products in the secondary market. They find that the profitability and environmental impact of leasing relative to selling depends greatly on the "durability" parameter δ and the product's environmental impact during the use stage of the life cycle. For IT equipment, where software is another significant feature of the product, OEMs can charge software relicensing fees for buyers of used or remanufactured equipment. Oraiopoulos *et al.* [10] study the case where an OEM can directly affect the resale value of its product in the secondary market through a mandatory software relicensing fee charged to the buyer of the remanufactured product, where remanufacturing is conducted by a 3PR. They find that it is suboptimal for the OEM to charge high relicensing fees, even though lower relicensing fees increase competition for the OEM's products by making the remanufactured product more attractive (i.e., lower priced). This occurs because consumers are strategic and consider the lower resale value for a product with high relicensing fees when purchasing a new product. Thus, a lower relicensing fee increases the demand for the OEM new product, and this market expansion effect again outweighs the cannibalization effect by the 3PR's remanufactured product.

Price as a quality signal: A key feature of the demand curves above is that demand for the remanufactured product decreases in its price, as shown in Figure 3.1(d).

Prices that are too low, however, can signal lower quality to consumers, as shown empirically by [11]. He conducts experiments to study the impact of prices of new and equivalent remanufactured laptops on the sales of remanufactured laptops—the cannibalization effect. He measures the cannibalization effect as the fraction of customers who switch from buying a new laptop to buying an equivalent remanufactured laptop for a given percent discount in the price of the remanufactured product relative to the new laptop's price. Figure 3.1(d) indicates that this fraction should be increasing in the remanufactured laptop's price discount, because the quantity of remanufactured products sold q_r should increase as the price of the remanufactured product p_r decreases (i.e., the discount for the remanufactured product increases). Ovchinnikov [11], however, finds the inverse U-shaped curve displayed in Figure 3.4, which means that the theory holds for discounts up to 20% in his case. For price discounts above 20% for remanufactured laptops, it appears that some consumers infer a lower quality of the remanufactured product, decreasing its demand. Thus, discounts above 20% can be detrimental to the profitability of

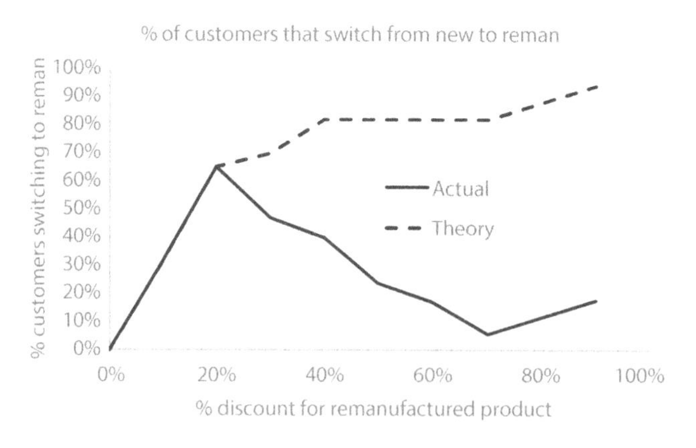

Figure 3.4 Price as a quality signal for remanufactured products.

remanufactured laptops. Evidently, this threshold is different for different product types, and different consumers such as industrial customers.

3.3 A Key Tactical Decision: Core Acquisition

Most tactical issues in CLSCs revolve around product (core) acquisition, and core disposition, where possible disposition decisions include remanufacturing, disassembly for spare parts, and materials recycling, among others. Remanufacturing planning has the peculiar characteristic of non-uniformity of inputs, as the quantity and quality of cores available for remanufacturing, the key inputs, in any given period can be uncertain. The firm has some control over the quality and quantity of cores, however, through its choice of acquisition prices. In addition, some firms have control over their demand for remanufactured products through its choice of remanufactured product price. In other words, acquisition prices (for cores) are a lever for controlling the supply side of remanufacturing, whereas the price of the remanufactured product controls the demand side of remanufacturing.

I present here decision models for the case where the firm's single disposition option in a period is remanufacturing (with recycling for materials recovery occurring when the core turns out to be unfit for remanufacturing). First, I present here a modified version of the framework of [8], who consider product acquisition and remanufacturing in a given planning period (say, one month), to illustrate the idea of matching supply of cores with demand for remanufactured products. In their model, which is based on cell phone remanufacturing, there are N quality categories (grades) for cores. For example, grades could be denoted by good better and best, and in this example $N = 3$. Grade i has unit remanufacturing cost c_{ri}, and remanufacturing yield y_i (if Q_i cores of grade i enter the remanufactured process, then only y_iQ_i are successfully remanufactured, with the remainder recycled for materials recovery.) For grade i, the firm sets its acquisition price a_i, and is able to procure $R_i(a_i)$ cores, where $R_i(\cdot)$ is increasing in a_i—the higher the acquisition price the firm offers for a given quality grade, the more cores it is able to procure. The firm's choice of remanufactured product price p_r, as suggested in the previous section, also determines its demand for remanufactured products in any given period. That is, the demand for the remanufactured product is $q_r(p_r)$, where the demand function (quantity sold) is denoted by $q_r(p_r)$. For example, we could have a linear demand curve as shown in Figure 3.1, so $q_r(p_r) = E - F p_r$.

Because the supply of returns, which is determined by a_i, constrains the amount of products that can be remanufactured, the firm should set its acquisition prices in a way that the amount of cores it is able to procure matches demand for remanufactured products, after accounting for remanufacturing yields in each of the grades: $q_r(p_r) = \sum_{i=1}^{N} y_i R_i(a_i)$. The firm's optimization problem is thus its choice of acquisition prices a_i for each core grade i that maximizes its profit in the period:

$$\max_{a_1,\ldots,a_n} \text{Profit} = p_r q_r\left(p_r\right) - \sum_{i=1}^{N} R_i\left(a_i\right)\cdot\left(a_i + c_{ri}\right) \quad (3.5)$$

$$\text{subject to } q_r\left(p_r\right) = \sum_{i=1}^{N} y_i R_i\left(a_i\right) \quad (3.6)$$

The problem defined by (3.5)-(3.6) would only apply to a scenario where cores can be objectively classified into different quality grades, and the firm has some leverage in setting the acquisition prices. The solution to (3.5)-(3.6) depends on the specific form of the core procurement functions $R_i(a_i)$, and the reader is referred to [8] for more details and solution. For some applications, core quality is more continuous, and acquisition prices are set competitively in the industry. I present now a product acquisition model for such a scenario.

Galbreth and Blackburn [6] also consider a single planning period, motivated by remanufacturing of toner cartridges. Here, the acquisition price per unit a is fixed (i.e., exogenous, not a decision variable), and demand D for remanufactured products is known in the period. Total remanufacturing cost, excluding acquisition cost for cores, decreases in the total quantity of returns acquired R, which is the decision variable. This is because by sourcing more cores than demand $R > D$, the firm can select only the best D cores to remanufacture, with the remainder $R - D$ being recycled for materials recovery. Thus, the firm wants to find the optimal acquisition quantity R for a desired demand D. We can translate this tactical decision to an acquisition *policy*, which is independent of the demand D, as a desired remanufacturing *yield* $y = D/R$. The quality condition of the cores can be modeled by a probability distribution of unit remanufacturing cost, denoted by $G(t)$ = Pr{Unit remanufacturing cost $\leq t$}. The curve is illustrated in Figure 3.5. The acquisition policy is shown in this curve: the maximum allowable unit remanufacturing cost t_{max} results into a probability of

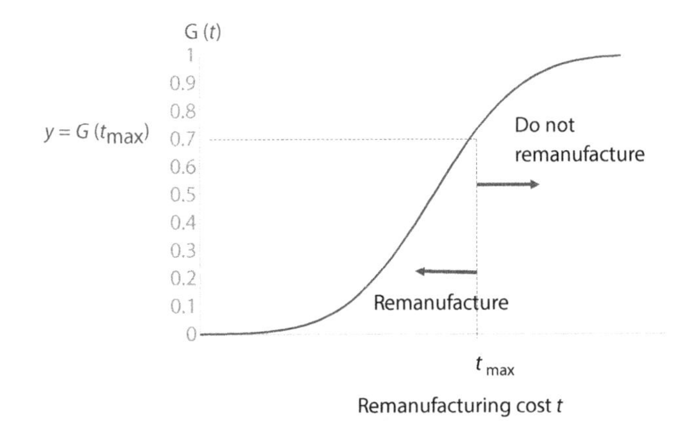

Figure 3.5 An acquisition policy selects the yield y that results in the maximum unit remanufacturing cost t_{max}.

$y = G(t_{max})$ that unit remanufacturing cost is less than t_{max}. Thus, if the firm sources R cores, it will result in $R \cdot y$ cores being remanufactured, which is precisely equal to the demand D. As a result, the firm's acquisition policy can be stated as finding the optimal yield y.

To find this optimal yield y^*, consider that the firm's total cost comprise (i) the acquisition cost, which is equal to the unit acquisition cost times the acquisition quantity, $a_y \cdot R$; and (ii) the total remanufacturing cost, which is equal to $R\int_0^y tg(t)dt$, where $g(t)$ is the probability density function of remanufacturing cost (i.e., g(t) = $dG(t)$/dt). Notice that $\int_0^y tg(t)dt$ is simply the average unit remanufacturing cost provided by the acquisition policy y.

Considering that $y = D/R$, then we substitute $R = D/y$ into the above cost components, so that we can express the firm's acquisition decision y as the solution to the following optimization problem:

$$\min_y \text{Total Cost} = a \cdot \frac{D}{y} + \frac{D}{y} \cdot \int_0^y tg(t)dt \quad (3.7)$$

$$\text{subject to } 0 \le y \le 1 \quad (3.8)$$

Notice that the demand parameter D multiplies the objective function (3.7), and thus it is simply a scaling factor of the total cost function. A s a result, the optimal acquisition policy y^* is independent of D, and it is only a function of the probability distribution of costs $G(t)$ and the unit acquisition cost a. This optimal policy can be found numerically through a line search in a spreadsheet. Simply put, the firm remanufactures the top y^*% of the cores it acquires. The trade-off here is better visualized in a chart if the total cost function is expressed in terms of the acquisition quantity R. Substituting $y = D/R$

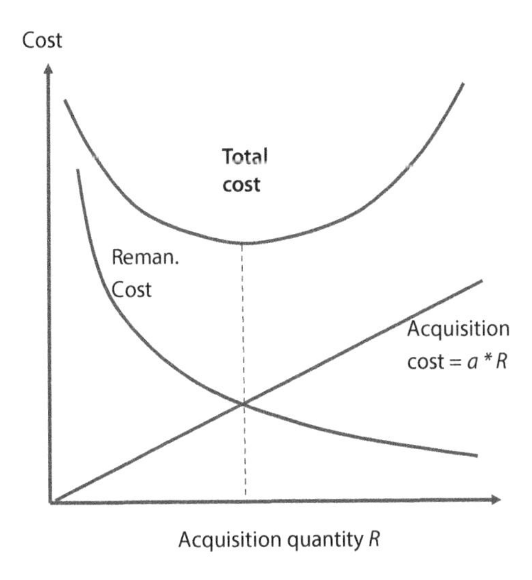

Figure 3.6 Trade-off in product acquisition for remanufacturing.

into (3.7) and plotting it, for a given set of parameters, we have Figure 3.6. The optimal acquisition quantity R minimizes the total cost curve. This problem has a structural similarity to the traditional economic order quantity (EOQ) model for finding the optimal order quantity, which makes it appealing, due to its simplicity

3.4 Conclusion

Remanufacturing is a key component of the circular economy, as it is a value-added activity that closes the loop, and recovers value in products post-consumer use. The environmental benefits of remanufacturing are typically clear, due to its significant savings in energy and materials compared to manufacturing a product from raw materials. There are cases, however, where recycling for materials recovery is the best option for closing the loop from an environmental perspective, particularly when the product consumes the bulk of its life cycle energy during use with consumers, as is the case with refrigerators, and internet routers. Because sustainability is concerned with economic, environmental, and social aspects, one needs to understand the business aspects of remanufacturing so that it becomes truly sustainable.

In this chapter, I have argued that two key decisions in remanufacturing, one strategic and one tactical, regards whether a firm should engage in remanufacturing in the first place, along with pricing of remanufactured products relative to corresponding new products, and policies for core acquisition. I show that an OEM, without significant threat of entry by third-party remanufacturers (3PR), introduces a remanufacturing product if the ratio of unit remanufacturing cost (including core acquisition) to unit new product cost is lower than a parameter δ, which measures the market acceptance of remanufactured products. The parameter δ can be readily estimated, and there are some estimates available for some products in some industries. The threshold ensures that the market expansion effect of offering a remanufactured product outweighs the cannibalization effect. When there is a significant threat of 3PRs, then the OEM remanufactures even for higher remanufacturing costs than the threshold, in order to fight competition by 3PRs.

Regarding product acquisition, I present two models from the remanufacturing literature that determine an optimal acquisition policy. In one model, the cores can be categorized into a finite number of quality grades, and the firm has the freedom to choose acquisition prices to determine the acquisition quantities, in each quality grade. The firm sets acquisition prices to match its demand for remanufactured products, balancing higher acquisition costs for higher quality grades against their lower remanufacturing costs. In the other model, quality grades are more continuous in nature, and acquisition prices are competitively dictated by the industry. In that case, the firm sets an acquisition policy that achieves a desired remanufacturing yield: by acquiring a larger quantity of cores than demand for remanufactured products, which increases the firm's acquisition cost, the firm can choose the better quality cores to remanufacture, decreasing its remanufacturing cost. The model balances these competing forces, and has a structural similarity with the familiar EOQ model for determining order quantities.

There are several other decision models available for remanufacturing, considering different industries, and different strategic, tactical and operational decisions than the ones discussed here. The reader is referred to [4] and [15] for reviews of this literature and further guidance.

References

1. Abbey, J.D. and Guide, V.D.R., Jr., Consumer markets in closed-loop supply chains, in: *Environmentally Responsible Supply Chains*, A. Atasu (Ed.), pp. 3–17, Springer, New York, 2016.
2. Agrawal, V., Ferguson, M., Toktay, L.B., Thomas, V., Is leasing greener than selling? *Manage. Sci.*, 58, 3, 523, 2012.
3. Atasu, A., Sarvary, M., Van Wassenhove, L.N., Remanufacturing as a marketing strategy. *Manage. Sci.*, 54, 10, 1731, 2008.
4. Ferguson, M. and Souza, G.C. (Eds.), *Closed-loop supply chains: New developments to improve the sustainability of business practices*, CRC Press, Boca Raton, 2010.
5. Ferguson, M. and Toktay, L.B., The effect of competition on recovery strategies. *Prod. Oper. Manag.*, 15, 3, 351, 2006.
6. Galbreth, M. and Blackburn, J.D., Optimal acquisition and sorting policies for remanufacturing. *Prod. Oper. Manag.*, 15, 3, 384, 2006.
7. Guide, V.D., Jr. and Li, K., The potential for cannibalization of new product sales by remanufactured products. *Decis. Sci.*, 41, 3, 547, 2010.
8. Guide, V.D., Jr., Teunter, R.H., Van Wassenhove, L.N., Matching demand and supply to maximize profits from remanufacturing. *Manufact. Serv. Oper. Manag.*, 5, 4, 303, 2003.
9. Hauser, W. and Lund, R.T., *The remanufacturing industry: Anatomy of a giant*, Boston University, Boston, 2003.
10. Oraiopoulos, N., Ferguson, M., Toktay, L.B., Relicensing as a secondary market strategy. *Manage. Sci.*, 58, 5, 1022, 2012.

11. Ovchinnikov, A., Revenue and cost management for remanufactured products. *Prod. Oper. Manag.*, 20, 6, 824, 2011.
12. Pochampally, K.K., Nukala, S., Gupta, S., *Strategic planning models for reverse and closed-loop supply chains*, CRC Press, Boca Raton, 2008.
13. Quariguasi-Frota-Neto, J. and Bloemhof, J., An analysis of the eco-efficiency of remanufactured personal computers and mobile phones. *Prod. Oper. Manag.*, 21, 1, 101–114, 2012.
14. Smith, V.M. and Keoleian, G.A., The value of remanufactured engines: Life-cycle environmental and economic perspectives. *J. Ind. Ecol.*, 8, 1–2, 193, 2004.
15. Souza, G.C., Closed-loop supply chains: A critical review, and future research. *Decis. Sci.*, 44, 1, 7, 2013.
16. Subramanian, R. and Subramanyam, R., Key factors in the market for remanufactured products. *Manufact. Serv. Oper. Manag.*, 14, 2, 315, 2012.
17. Tereyăgŏglu, N., Market behavior towards remanufactured products, in: *Environmentally Responsible Supply Chains*, A. Atasu (Ed.), pp. 19–28, Springer, New York, 2016.

4

Remanufacturing, Closed-Loop Systems and Reverse Logistics

Rolf Steinhilper* **and Steffen Butzer**

Chair Manufacturing and Remanufacturing Technology, University of Bayreuth, Bayreuth, Germany

Abstract

This chapter provides an overview of the concepts of Remanufacturing, Closed-Loop Systems, and Reverse Logistics, as well as the connections between them. The reasons for Reverse Logistics and remanufacturing, the role of cores in each concept, core return strategies, sourcing models, and both barriers to and drivers of implementation are also discussed.

Keywords: Remanufacturing, closed-loop supply chain, reverse logistics

4.1 Introduction

Today, manufacturing companies are faced with challenges that are twofold and seemingly at opposition with one another. On one hand, they must balance demands for mass customization with shorter product life cycles and increasing global competition; on the other, they are driven to care about the increasing customer awareness of environmental impacts and

**Corresponding author*: rolf.steinhilper@uni-bayreuth.de

Nabil Nasr (ed.) Remanufacturing in the Circular Economy (85–110)

the relative "eco-friendliness" of their products, processes, and systems.

An emerging approach through which to address these challenges is the Circular Economy Movement, a socio-industrial construction whereby the outputs of one process or lifecycle phase reciprocally become inputs in another. This concept is highlighted globally by scholars, policymakers, and industrialists who believe economic growth can be achieved under current conditions without increased resource dependence. In these pursuits, the concepts of Closed-Loop Systems, Reverse Logistics, and Remanufacturing are vital to success.

However, identifying and implementing suitable industrial models in each of these areas remains a challenge. To this end, the following chapter aims to provide an overview of the concepts of Remanufacturing, Closed-Loop Systems, and Reverse Logistics, as well as the connections between them that must be strengthened in order to support a circular economy.

4.2 Remanufacturing in Closed-Loop Systems

In an era of globally-connected production networks, supply chain management is a key discipline of modern manufacturing and consumption patterns. The classical supply chain concept, i.e. traditional forward logistics, is well established and serves as the foundation for most of today's global manufacturing activities. Within this model, manufacturers source raw materials or semi-finished goods from suppliers and add value to them through subsequent manufacturing processes. Following manufacturing, distributors sell finished goods to retailers in bulk, who then resell the goods to customers.

While these classical supply chain models are effective, they lack in efficiency, and ultimately create issues in stability with respect to inherently finite global reserves of new material. To address these concerns, the following section explores the closed loop supply chain model and, at a high-level, how remanufacturing can serve as a strategy in creating closed-loop systems.

4.2.1 Closed-Loop Supply Chains and Systems

The product lifecycle does not have to end with delivery to the end-customers [2]. In many cases, products or the materials within them can be recovered and restored to serve additional lifecycles, reducing the raw material required to produce new replacements. To accomplish this, products at the end of their life (EOL) or use cycle can be collected by retailers, distributors, manufacturers, or suppliers for repair, reuse, remanufacturing, or material recycling.

This reverse flow of goods backwards through the elements of a classical supply chain is called, aptly, Reverse Supply Chain (RSC), and depends accordingly upon Reverse Logistics (RL) strategies. In these systems, EOL products that are taken back are called cores. Applying RSC techniques to a classical forward supply chain with an iterative, multi-step model can circularize the flow of products and materials, creating a closed-loop supply chain that is cyclically regenerative. Combined with the operations performed at each process step (e.g. manufacturing and remanufacturing), a closed-loop supply chain becomes a Closed-Loop System.

Figure 4.1 illustrates such a Closed-Loop System, in which used or returned products - cores are collected, inspected, sorted into different categories, and used in repair, reuse, remanufacturing, recycling, or disposal, depending on their classification [1].

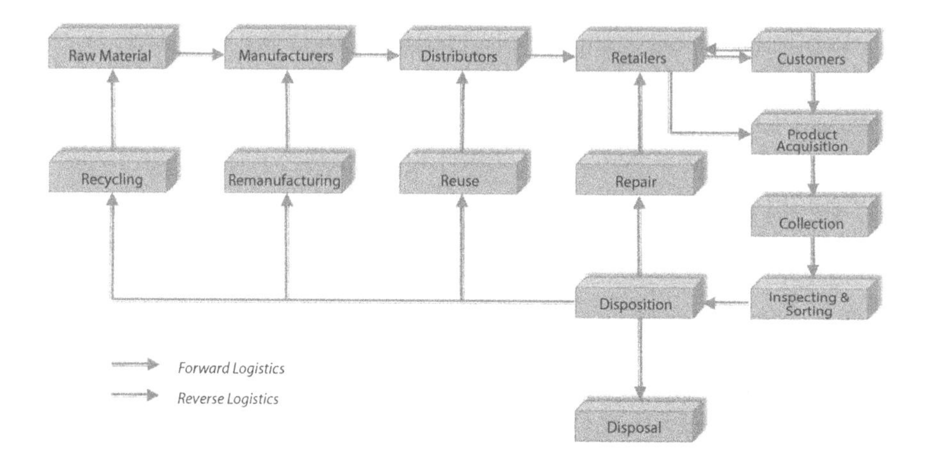

Figure 4.1 The closed-loop supply chain according to Agrawal *et al.* [1].

4.2.2 Differentiation of Regeneration Approaches

The differences between the regeneration approaches in Figure 4.1 are explained in the following.

Repair

Products that are broken can be repaired and directly reincorporated into retail markets. Within the repair process, only defective parts and components are exchanged, while functioning parts are not treated. Repaired products may thus be considered of lower value than new products. These products remain within their first life cycle, and in many cases, their first use cycle. Examples include mobile phone displays or washing machine pumps.

Reuse

Some noncomplex products can be directly reused by customers without or with only minor inspection, cleaning, and maintenance [1]. Examples include glass bottles or food boxes.

Remanufacturing

Remanufacturing is the process used to restore EOL products to new-equivalent condition or better. What separates remanufacturing from the approaches outlined above is the degree of effort expended to restore products. While repair and reuse aim to extract whatever value remains in a first-lifecycle product, remanufacturing aims to recover the value embedded within the quality and technology of its components to create another lifecycle altogether. The number and sequence of process steps—and thus the cost-benefit balance of remanufacturing—depend on the type and functionality of the product. In general, remanufacturing requires five common process steps: disassembly, cleaning, inspection, reconditioning, and reassembly Electronics-rich products, however, often require a sixth preceding step—entrance diagnosis—in order to determine the cause of non-mechanical failure and the best course of subsequent action (Figure 4.2).

Recycling

The purpose of recycling is to recover only the value of materials from returned products, rather than the value of its final form or functionality. Products used in recycling are typically crushed or shredded, and their constituent materials separated and smelted. While recycling reduces the costs for disposal

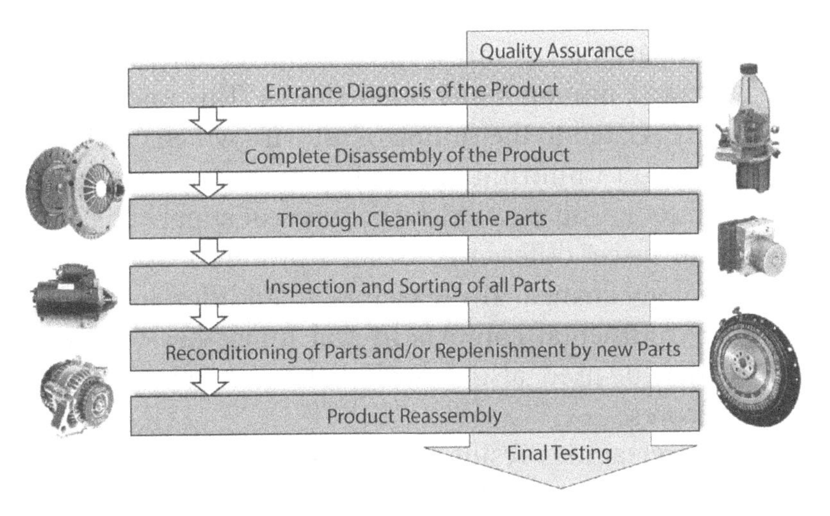

Figure 4.2 Primary remanufacturing process steps.

[3], it is the least efficient and value-preserving approach to regeneration.

In terms of economic potential, remanufacturing preserves the value-added in new production and allows it to be extracted multiple times across several life cycles. Through this preservation, remanufacturing can help avoid the material and energy costs of new production. Ecologically, this also leads to corresponding resource savings, avoids emissions, and reduces material waste.

4.2.3 The Role of Cores for Remanufacturing

Remanufacturing of course, is not a new strategy; it has been employed for decades as a means to address material scarcity and capital-intensive product maintenance. However, in order to implement remanufacturing as a central industrial practice from which sustainable profit can be derived, it is crucial that remanufacturing companies match the demand for remanufactured products with the supply of available cores. This is not an easy task; factors beyond only quantity—such as timing, quality, pricing, and product specifications—must all be considered.

The quantity of available cores is made uncertain by the diversity of customers' environmental awareness, the dynamic market competition for cores, and the convenience of logistics infrastructures, among many other things. This uncertainty is also directly related to the timing of returns, which can itself be affected the non-uniformity of the life cycle stage at which any given product group may exist, the types of supply chain relationships in place, differentiation in a product's usage period, and technology change. These factors, in addition to the diversity of environmental contexts in and intensities with which products may be used, also contribute to uncertainty about the quality of cores.

These uncertainties result in a mismatch of product return (i.e. supply) and demand. On the one hand, if not enough quality cores are returned, remanufacturers must salvage cores with lower quality, convert other types of cores, or even use new products to meet demand. On the other hand, an overstock of cores increases inventory costs and the risk of obsolescence.

Ultimately, the above-mentioned uncertainties cause complexity in resource planning, increase uncertainties in processing times, and create unnecessary effort for controlling remanufacturing operations [4]. Figure 4.3 illustrates these common supply and demand conjunction scenarios.

Though expensive, overstocking can guarantee core availability to some degree; however, this strategy is not viable in the long-term. In cases of chronic supply bottlenecks, relying on initial overstock will subject the remanufacturer to price shocks and an unpreparedness to sustain operations once the overstock is diminished. Conversely, for long-duration remanufacturing processes, overstocking of inventory creates risk that storage costs will increase exponentially without any clear means to reduce that burden. To address these concerns, Privono *et al.* [5] suggest strategies for better coordination of demand and supply, which are summarized in Table 4.1.

Contractual agreements between suppliers, customers and remanufacturing companies are useful in ensuring a stable

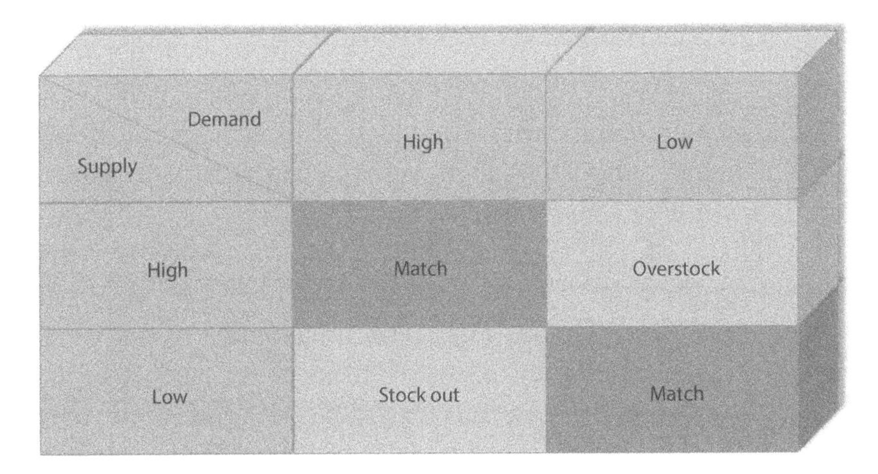

Figure 4.3 Matching demand and supply, according to Privono *et al.* [5].

Table 4.1 Strategies to match demand and supply of cores, according to Privono *et al.* [5].

Supply side (Product Acquisition Management)	**"Remanufacturing Engine" Reman Operations**	**Demand side (Remarketing Issues)**
Establish contracts with suppliers and customers (OEMs)	Production leveling and build stock in used products to maintain operation	Building contracts with OEMs; 1% warranty sales
Information System for information sharing and drafting production plan	Flexible workers and work cells production layout	Forecasting based on historical data, new product sales, and collaborative forecasting
Forecasting based on historical data, new product sales, and collaborative forecasting organised by a commercial team	New parts stocks	Finished product stock and using "dummy components" for out stock parts

balance in supply and demand. Likewise, forecasting and sharing information about production plans can reduce variations that increase economic risks.

The effect of core availability on the necessity for new (virgin) product parts and components is also a central issue of core supply and demand. In accordance with Privono *et al.*, Figure 4.4 illustrates that as the volume of returned product cores decreases, the availability of valuable replacement parts decreases, subsequently creating an increased need to integrate new parts in remanufacturing.

This relationship can have significant impacts on both the economic and environmental profiles of remanufacturing. The profitably of remanufacturing is naturally limited by the availability and relative value of cores and spare parts. Thus, profitability ends when the value of required new exceeds that which can be recovered from used cores. As Figure 4.4 illustrates, this can be a relatively narrow area; however, continuing remanufacturing beyond this point can be justified temporarily if a high demand is expected to continue to a point where core availability is projected to increase again.

Profitability also depends upon the costs (i.e. intensity) of required remanufacturing effort and the sales price that can be achieved. In general, lower-quality cores hold lower value and thus achieve lower the expected sales price and require higher remanufacturing effort. In this sense, profitability is in large part also determined by core quality, and variability in quality thus affects the stability of profitability and, ultimately, a company's financial performance [5].

Beyond these factors, other approaches for product regeneration (e.g. reuse and repair) can also significantly affect the demand for remanufactured products [6]. Resale, for example, can be a more viable solution in secondary markets if a product is in working condition and its expected remaining lifetime is

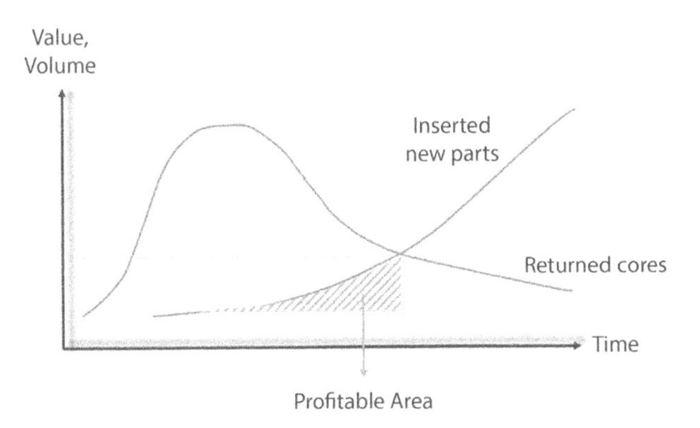

Figure 4.4 Trade-off between parts rejection and new parts insertion, acccording to Privono *et al.* [5].

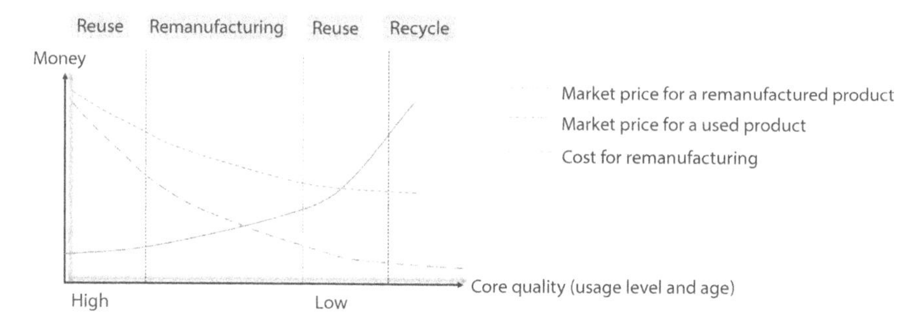

Figure 4.5 Economically-preferable approach to product regeneration as a function of core quality, adapted from forklift truck case [6].

comparable to other products in its family. In practice, this decision is largely a function of the cost of each option, as illustrated in Figure 4.5. For example, when the quality of a first-lifecycle forklift truck is high, there is no motivation to remanufacture because its value on a secondary market is sufficient, and the cost for remanufacturing is greater than the potential increase in achievable market price. Later, when the forklift is worn, the resale price of a remanufactured product is higher because demand has shifted away from first-use resale products that no longer meet customer expectations for quality and remaining life. Thus, the cost for remanufacturing is less than the additional value that may be gained. This requires a balance, however; for cores with low quality, the cost of remanufacturing can become too high and thus exceed the possible value on a secondary market. In this situation, the used product can instead be resold to a market segment with lower quality demands, such as a customer with limited or intermittent need for a forklift. For this customer, a lower-quality forklift is appropriate because the level of use is so low that the remaining life will be sufficient.

4.3 Reverse Logistics

Several industrial processes are available through which classical supply chain systems may be developed to become more closed-loop, and therein more resource-efficient. Ensuring that these processes can be implemented and utilized on a broad industry scale, however, requires economically feasible strategies to encourage and accomplish the collection and appropriate redistribution of end-of-use and EOL products and materials. This "process of planning, implementing, and controlling the efficient, cost effective flow of raw materials, in-process inventory, finished goods, and related information from the point of consumption to the point of origin for the purpose of recapturing value" is known as Reverse Logistics [7].

4.3.1 Justifications for Reverse Logistics and Remanufacturing

The following four themes, which companies cite as justification to engage in reverse logistics and remanufacturing, are widely spread across scientific literature: economics, legislation, corporate citizenship, and environmental responsibility. Each of these is discussed below.

Economics

Many companies use remanufacturing as a supplementary platform through which to save costs and to serve customers well. However, because remanufactured products perform as well as—and, in some cases, better than—their virgin counterparts, there is high potential to reach strong market penetration by meeting new and emerging customer demands. The high quality and attractive price-performance ratio of remanufactured products enhance brand appeal.

Perhaps greatest amongst these four themes, economic targets in cost reduction, value addition, and market share increase are often the prime mover of Reverse Logistics [8].

Legislation

Legislation is another important justification for Reverse Logistics and remanufacturing. In some cases, legislation can require companies to take EOL products back to recycle or even remanufacture them. Legislative efforts may govern the collection and reuse of EOL products, shift waste management responsibilities to producers, define allowable volumes of waste generation, require increased use recycled materials, and restrict the use and management of hazardous substances. Ultimately, legislation encourages companies to employ reverse logistics by creating business risks in the form of penalties and reputational effects [8]. This incentivizes companies to ensure that the EOL products are collected and processed in a way that is compliant with existing legislation [9].

Corporate Citizenship

The concept of corporate citizenship considers the corporation as a member of the society in which it exists, and seeks to qualify its civic performance. In other words, corporate citizenship measures and reflects the degree to which a company recognizes the impacts of its operations and its products, and the ways in which it benefits its community—or at least limits its degradation of that community—in ways other than simply creating and selling products. Often, public perceptions of corporate citizenship can dramatically affect a company's economic vitality. In this respect, companies are sometimes encouraged to create reverse logistics systems and engage in remanufacturing in effort to demonstrate their understanding of and take responsibility for the impacts of their products and operations, thereby leveraging the value customer populations place in corporate citizenship as a means to improve market recognition.

Environmental Responsibility

Customer perceptions of a company's value are also increasingly tied to environmental performance. As a result, companies are largely drawn to remanufacturing and reverse logistics as a means to capitalize upon the high savings potential in energy and raw materials. Both give companies the opportunity to promote their products as "green" and environmentally friendly.

4.3.2 Core Return Strategies

While these four key themes are widely cited in corporate justifications for developing reverse logistics systems and remanufacturing capabilities, their benefits cannot be leveraged, nor their risks avoided, without viable strategies to acquire product cores. This section describes several common core acquisition strategies used by Original Equipment Manufacturers (OEMs) and Independent Remanufacturers (IR) alike.

Deposit Fee

By requiring a deposit fee at the first sale, the customer receives a pecuniary incentive to return the product after its usage to a dealer or manufacturer. The advantage in this strategy is that the customer returns the core independently, requiring no further influence of any actor, and thus no further cost in processing. There is, however, a challenge in the difficulty and duration associated with creating market acceptance of such a deposit model.

Service Contract

Instead of selling a product, making it the sole responsibility of the customer, a service contract between a manufacturer and a customer may include remanufacturing as a means to benefit both parties—the customer is assured maintenance, while the producer saves costs [10].

Direct-Order

Customers may take an EOL product directly to a remanufacturer specifically for the purpose of having it remanufactured. Remanufacturers must decide the value of each case [10].

Credit-Based

Customers may return an EOL product in order to receive some form of credit, which may then be used toward the purchase of a replacement remanufactured product [10].

Buyback-Based

Producers may offer to repurchase a product from customers when it meets certain criteria in the future as a part of the initial product transaction [11]. The benefits of this are similar to those of the deposit model; however, buy-back programs require experts to develop criteria for the optimal time and condition for repurchase—an uncertain, difficult, and expensive endeavor.

Exchange System/Deposit-Based

By combining the deposit model with the credit model, remanufacturers can effectively oblige customers to act as suppliers [11]. In other words, if the remanufacturer requires customers to provide an EOL product core as a prerequisite to purchasing a replacement remanufactured product, this one-for-one exchange can create a steady, predictable stream of cores [10].

While this model may incentivize customers to purchase remanufactured products on an economic basis, a one-for-one exchange purchase prerequisite inhibits the customer's ability to purchase in volume and at will, as their own supply of EOL products may not keep pace with their demand for additional products.

Takeback with or without Costs for Supplier/Voluntary-Based

Suppliers, usually either customers or other producers [10], may voluntarily return EOL products to remanufacturers as a means of responsible disposal. In this model, the benefit to suppliers is that they are able to avoid storage or high disposal costs from other conventional disposition pathways, while remanufacturers effectively secure no-cost cores [11]. In cases where conventional disposal may be subject to high environmental costs, remanufacturers may even charge a fee for accepting cores to offset additional costs they may incur in processing.

Contractual Agreements

In this case, collectors and remanufacturers cooperate closely to negotiate and contractually predetermine the type, volume, and price of products the remanufacturer will purchase from the supplier. This ensures a degree of security in the sales business for collectors while providing a reliable core supply stream for the remanufacturer. Automobile junkyards and automotive parts remanufacturers use this type of acquisition strategy.

Lease or Rent Contracts/Ownership-Based

Products are not sold, but leased or rented, where the (initial) duration and end date of the contract is predetermined. Usually, the configuration and condition of products that are leased are quite well known throughout the contract, meaning the core knowledge upon return is high. The condition of the product upon return is influenced by the duration of the contract [11].

Central Returns Center

A central returns center (CRC) is a distribution center where returns of multiple origins flow back to a central collection point. When disposition decisions are centralized, large volumes of each product can be processed, allowing technical personnel to learn more quickly and develop experiences regarding best practices and revenue-maximizing techniques. Companies that purchase these products prefer to deal in large volumes, and the concentration of goods in one place appeals to potential bidders, increasing their willingness to pay. The major perceived disadvantage of CRCs is the potentially high transport costs and the risk that the product may ultimately be thrown away [7].

Legislation

Imposing high disposal costs or enforcing outright bans on conventional disposal of certain products inherently creates demand for a more cost-effective means of disposition. When core collectors can meet this demand by accepting EOL products, they can secure a low-cost supply stream from the user population [11]. The potential for success of this strategy is highly dependent on specific laws and regulations, which vary greatly by country, region, and market philosophy. As a result, no single legislative solution may address challenges globally.

According to Sundin *et al.*, the two most common sourcing strategies are ownership-based and buyback-based sourcing. In many cases, companies use a combination of several strategies, often supplementing the primary strategy with voluntary-based options [12].

4.3.3 Barriers of Reverse Logistics and Remanufacturing

According to Rogers & Tibben-Lembke, many companies have difficulties in executing reverse logistics due to internal and external barries [13]. Table 4.2 outlines the severity and centrality of these issues as found by Rogers & Tibben-Lembke in

Table 4.2 Barriers to Reverse Logistics systems, according to Rogers & Tibben-Lembke [13].

Barrier	Relevance in percentage of cases
Low importance of reverse logistics relative to others issues	39.2%
Company policies	35%
Lack of systems	34.3%
Competitive issues	33.7%
Management inattention	26.8%
Lack of financial resources	19.0%
Lack of personnel resources	19.0%
Legal issues	14.1%

a survey of over 300 companies. These barriers are described in more detail in the following subsections.

Relative Importance

According to the study, the perceived unimportance of reverse logistics compared to other business issues is reported as the single most significant barrier for adoption in industry. It is important to consider that since the time of publishing (1998) the exposure to and emergence of remanufacturing and reverse logistics within the industrial business framework has increased. The resultant industrial sensitization to these concepts may have reduced this barrier somewhat in the contemporary context; nevertheless, these concepts have not yet gained solid acceptance in all areas of industry, and the perception of their relative importance therefore remains a significant hurdle that must be overcome.

Company Policies

Many manufacturing companies suggest a reluctance to engage in remanufacturing rooted in concern that remanufactured products will cannibalize new products sales. This often leads to restrictive company policies and product design choices explicitly intended to prevent remanufacturing. This conceptual barrier remains one of the most difficult to overcome, as the economic and ecological advantages of reverse logistics and remanufacturing are, for many, difficult to grasp in the context of a classical linear industrial paradigm.

Lack of Management Systems

Because the field of reverse logistics is still relatively young, few well-functioning management strategies have been successfully developed or deployed. Such systems are a critical foundation for adoption, and must be developed within the near future.

Competitive Issues

Declining competitiveness reduces the willingness to engage in reverse logistics and closed-loop supply chains. This barrier reflects economic shortsightedness, however, as the technology and knowledge in remanufacturing today supports these systems as viable economic strategies.

Management Inattention & Resource Allocation

Because the present upper management paradigm often dismisses reverse logistics, closed-loop systems, and remanufacturing as secondary or noncompetitive business practices, the financial and human resources that are then allocated to these concepts are insufficient to create systems that demonstrate otherwise. This ultimately creates a self-fulfilling prophecy of non-viability that is in many cases cited as further justification for a lack of investment.

Legal Issues

Contrary to some widely-held perspectives, legislation appears to be the least-inhibitory influence considering its position in this ranking of barriers [13].

4.3.4 Drivers of Reverse Logistics and Remanufacturing

In contrast to these barriers, there are also many factors that indeed drive companies to engage in reverse logistics and remanufacturing [14]. These drivers are described below.

Government Support

Through the support of government policies and legislation, financial incentives may be created which facilitate and expand the adoption of reverse logistics and remanufacturing systems.

Cost Reduction

By using previously manufactured goods in remanufacturing, the incurred material, energy, and process costs—as well as the associated environmental impacts of each—are substantially lower than in new production. This not only reduces direct process costs, but also mitigates environmental costs, waste management and disposal fees, and material and regulatory risks.

Environmental Protection
As described above, adopting closed-loop systems and remanufacturing practices can save material, energy, and emissions [12], thereby reducing the industry's impact on the environment. Importantly, this preserves resources for future generations to sustain business.

Corporate Image
Reducing both costs and environmental impacts creates a public perception of a positive, sustainable corporate ethos. This is known to correlate with an increased acceptance in society and amongst customers, ultimately improving business opportunity.

Improved Customer Service
By engaging in remanufacturing (and thus inherently in reverse logistics), products and spare parts and products that are no longer produced new or prohibitively difficult to acquire can still be provided to meet continuing customer demand. This allows for a longer product use cycle at low storage and maintenance costs for the customers improving their satisfaction.

4.3.5 In- or Outsourced Reverse Logistics

Following the decision to engage in remanufacturing, companies must decide whether closed-loop reverse logistics systems can be built with existing infrastructures, or must be outsourced to a third party. Figure 4.6 illustrates this decision as well as three proposed management models.

Each of these three models for reverse logistics management are described below:

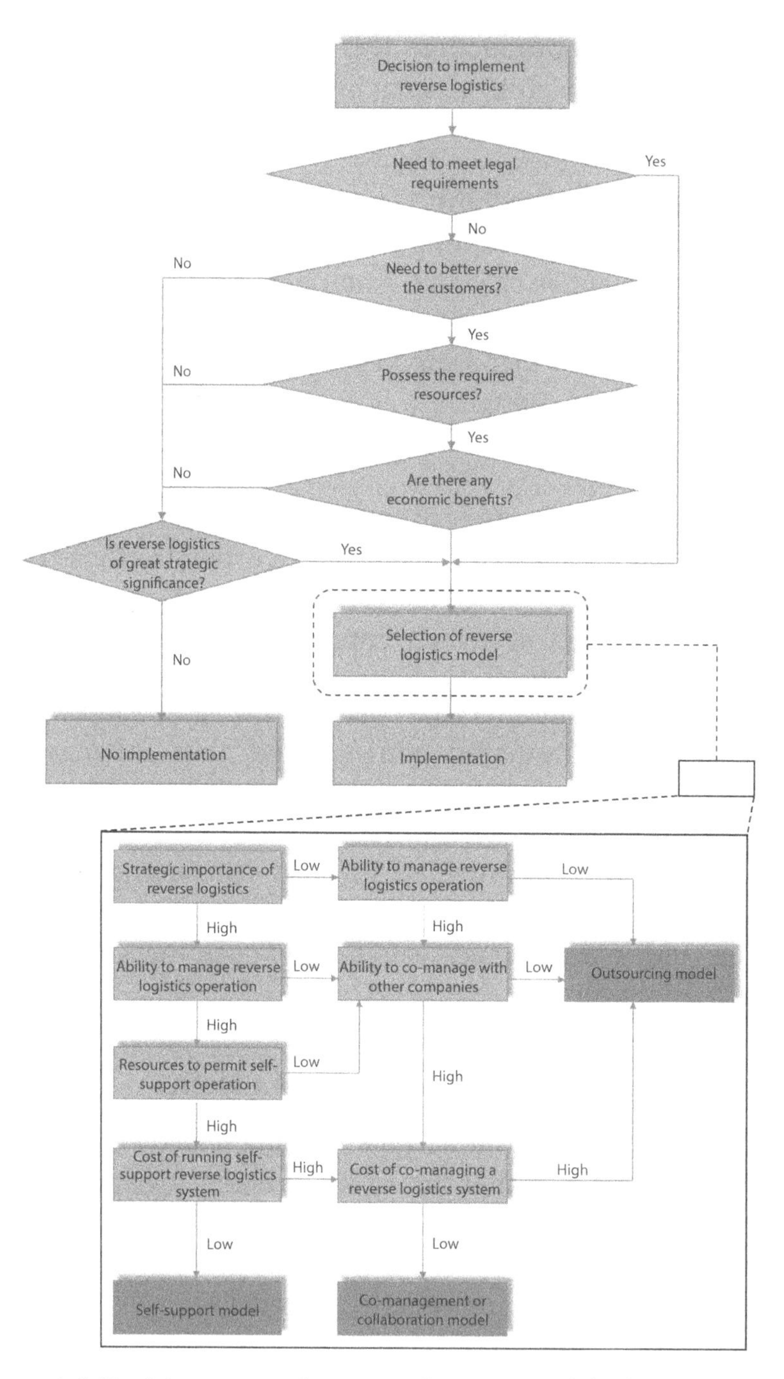

Figure 4.6 Decision process for reverse logistics model selection and implementation, according to Lau & Want [14].

Self-Support

In this model, the manufacturer/remanufacturer develops its own reverse logistics system by leveraging existing supply and distribution channels and creating a unique management framework to address circular product and material flows. In this case, cores may be sourced from the same distributors, retailers and/or end users that make up the classical forward supply chain. This model prevents the company's product knowledge from seeping outward, making it difficult for competitors to remanufacture their products, and thus protects market position.

Third-Party

In this model, core recovery and reverse logistics are performed by an external company and cores are eventually shipped to the producer for remanufacturing. This model enables the manufacturer/remanufacturer to focus on remanufacturing operations and protect their competitiveness; while the supply chain is known to a third-party, product specifications and remanufacturing process expertise can remain proprietary.

Collaborative

In this model, multiple manufacturers and/or remanufacturers collaborate to create a mutually beneficial system for collective core recovery and reverse logistics. Synergies between the partners are created and additional knowledge is generated, serving the interests of all parties. While collaboration is beneficial, many companies are skeptical of adopting such practices out of competitive fear that collaborators may exploit this trust to gain an advantage.

Figure 4.7 illustrates these three reverse logistics models.

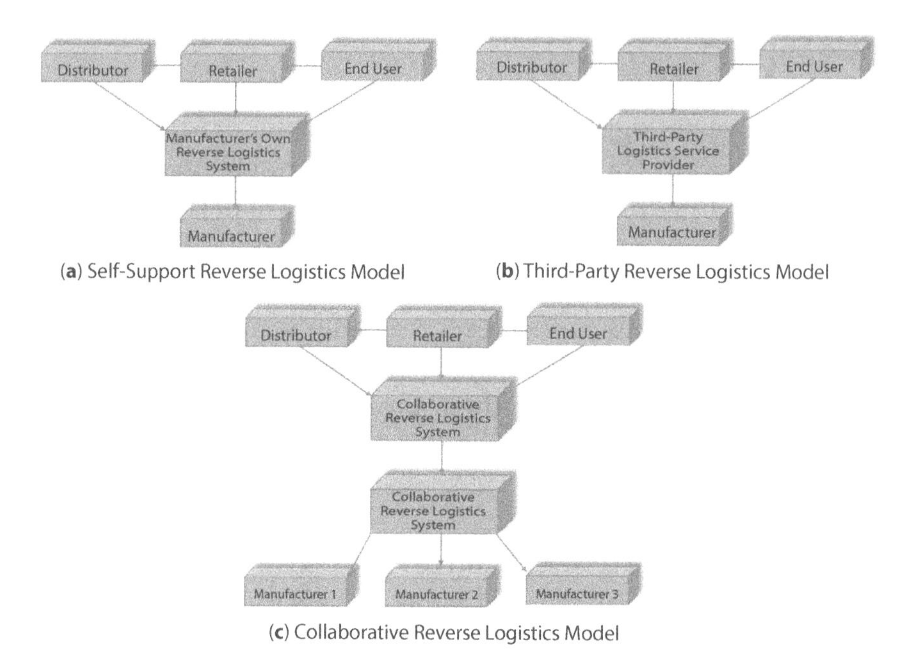

Figure 4.7 Return collection arrangement within different reverse logistics models, according to Lau & Want [14].

4.4 The Future of Reverse Logistics and Remanufacturing

The potential for closed-loop systems that leverage reverse logistics and remanufacturing to create significant environmental and economic benefits is well demonstrated on the theoretical model. However, much progress stands to be made in transitioning from current linear industrial and economic models towards more regenerative circular systems. Ultimately, the key to precipitating a shift in the industrial paradigm is to make clear that relying on customers, end-users, and consumers as a critical point in the reverse supply chain is and will continue to be both a reliable and economically viable structure. To this end, several factors—such as increasing global competition, shorter life cycles, environmentally-oriented legislation, and more lenient commercial takeback regulations—will act as mechanisms to increase product returns.

Of course, the balance of quality with quantity and the matching of supply to demand are paramount in this endeavor. Because quality and timing can affect economic margins in global markets even with increased returns, such a transition is not without risk [15]. Effective market and logistics planning, however, will do well to minimize this risk.

Part of this planning must be a broadening of corporate perspectives to consider the holistic lifecycles of products, process, and industrial systems. Only by looking at the complete lifecycle of a product—from conception through delivery to use and back again—can manufacturers and remanufacturers truly understand how existing business and technical approaches must be adjusted and strengthened to make closed-loop systems technically possible, environmentally preferable, and economically profitable. In this, considering the necessity for product returns (commercial, warranty, repairs, end-of-use, end-of-life, and the like) from the very beginning of business case development is absolutely critical, as companies must not only design their products, processes, and systems to function in the classical way, but must also engineer them to be successful in reverse [15].

Ultimately, the industrial economy will eventually be left no choice but to develop toward fully-integrated, closed-loop systems if it wishes to be sustained. In light of the conflicting challenges created by massive population growth, finite resources, and the desire for widespread economic development, circularity will become necessary. Adopting a transformational approach now is thus the only way to ensure preparedness, resilience, and systemic maturity when that point is finally reached.

References

1. Agrawal, S., Singh, R.K., Murtaza, Q., *A literature review and perspectives in reverse logistics*, pp. 77–92, Mechanical, Production & Industrial Engineering Department, Delhi Technological University, Delhi, India, 2015.

2. Shaik, M.N., *Comprehensive Performance Measurement Methodology for Reverse Logistics Enterprise*, pp. 1–233, University of Windsor, Windsor, Canada, 2014.
3. Chan, F.T.S., Chan, H.K., Jain, V., A framework of reverse logistics for the automobile industry. *Int. J. Prod. Res.*, 50, 5, 1318–1331, Department of Industrial and Systems Engineering, The Hong Kong Polytechnic University, Hong Kong, China, 2012.
4. Wei, S., Tang, O., Sundin, E., Core (product) Acquisition Management for remanufacturing: A review. *J. Remanufacturing*, 5, 4, 1–27. doi: 10.1186/s13243-015-0014-7, Springer, Division of Production Economics, Department of Management and Engineering, Linköping University, Linköping, Sweden, 2015.
5. Priyono, A., Bititci, U.S., Ijomah, W.I., *Balancing Supply and Demand in Reverse Supply Chain: A Case Study in Remanufacturing Company*, pp. 707–713, Department of Design, Manufacture and Engineering Management, University of Strathclyde, United Kingdom, 2012.
6. Östlin, J., Sundin, E., Björkman, M., Product Lifecycle Implications for Remanufacturing Strategies. *J. Clean. Prod.*, 17, 11, 999–1009, 2009.
7. Rogers, D.S. and Tibben-Lembke, R.S., An Examination of Reverse Logistics Practices. *J. Bus. Logist.*, 22, 2, pp. 129–148, University of Nevada, Reno, U.S.A., 2001.
8. Jun, W., *A Fuzzy Evaluation Model of the Performance Evaluation for the Reverse Logistics Management*, pp. 724–727, Zhongkai University of Agriculture and Technology, Guangzhou, China, 2009.
9. Ravi, V., Shankara, R., Tiwari, M.K., *Analyzing alternatives in reverse logistics for end-of-life computers: ANP and balanced scorecard approach*, pp. 327–356, Department of Management Studies, Indian Institute of Technology Delhi, Hauz Khas, New Delhi, India, 2005.
10. Östlin, J., Sundin, E., Björkman, M., Importance of closed-loop supply chain relationship for product remanufacturing. *Int. J. Prod. Econ.*, 115, 2, 336–348, 2008.
11. de Brito, M.P., Flapper, S.D.P., Dekker, R., *Reverse Logistics: A review of case studies*, pp. 1–32, Econometric Institute Erasmus University Rotterdam, Rotterdam, Netherlands, 2002.

12. Sundin, E., Sakao, T., Lindahl, M., Kao, C., Joungerious, B., Map of Remanufacturing Business Model Landscape, European Remanufacturing Network, For Horizon 2020, grant agreement No 645984, February 2016.
13. Rogers, D.S. and Tibben-Lembke, R.S., *Going Backwards: Reverse Logistics Trends and Practices*, pp. 1–275, Center for Logistics Management, University of Nevada, Reno, U.S.A., 1998.
14. Lau, K.H. and Wang, Y., Reverse logistics in the electronic industry of China: A case study. *Supply Chain Manag.*, 14, 6, 447–465, Logistics Group, School of Management, Royal Melbourne Institute of Technology University, Melbourne, Australia, 2009.
15. Guide, D., Harrison, T., Van Wassenhove, L., *The Challenge of Closed-Loop Supply Chains*, pp. 1–8, INSEAD Fontainebleau, France, 2003.

5

Product Service and Remanufacturing

Mitsutaka Matsumoto*

Advanced Manufacturing Research Institute (AMRI), National Insitute of Advanced Industrial Science & Technology (AIST) Tsukuba, Japan

Abstract

Remanufacturing offers significant potential benefits for industries and society, but has encountered several significant barriers in achieving market penetration and global uptake. This chapter argues that these barriers can be mitigated when remanufacturing is incorporated into the service business models of manufacturing companies. Currently, many manufacturing companies, especially those based in developed countries, are making efforts to add creative product services to their business models. This movement provides an opportunity to not only increase economic profits for manufacturers, but also to foster the development of remanufacturing in pursuit of a more sustainable industrial economy. This chapter presents the current state, possibilities, and challenges in product services and remanufacturing, and explores the potential to incorporate these two processes, highlighting case studies in two product areas to support this discussion. Finally, this chapter discusses requirements and challenges to effectively integrating product services and remanufacturing.

Email: matsumoto-mi@aist.go.jp

The previous study reviewing parts of this chapter are based on a study presented by the author of this chapter [1].

Nabil Nasr (ed.) Remanufacturing in the Circular Economy (111–136)

Keywords: Product services, servitization, product-service systems, after-sales service, function-selling, motivation for remanufacturing

5.1 Introduction

Contemporary global economic systems almost invariably rely upon mass production, mass consumption, and mass disposal; a dependence whose future is constrained, even threatened, by the inherently limited availability of material resources. In order to establish a more sustainable society, it is becoming increasingly imperative to consider products in the context of their entire lifecycle and thus close the loop with regard to material flows in product or service delivery. One element in this product lifecycle strategy that can help achieve this goal—creating numerous benefits both for industry and society—is remanufacturing. This process has, to some degree, been adopted globally in various sectors [2, 3], garnering attention for its potential benefits [4], but not necessarily the traction required to precipitate large-scale transition towards a more sustainable industrial economy. Though many technical and economic challenges still remain as barriers to growth in remanufacturing, efforts to increase recognition of its benefits and improve global market uptake are both necessary and growing.

Independent of these movements, but emerging concurrently, many manufacturing companies are making efforts to add services (e.g. maintenance) to their business portfolios, with this "servitization" supplementing sales-volume-based business models with services as revenue generators [5]. In an era where technological development in products and production systems alike has led to market saturation and commoditization in several product areas, many manufacturing companies—especially those based in developed countries—servitize their businesses as a means to both maintain and increase profits. In some cases, companies employ traditional revenue-generating services such as product maintenance and repair. However, innovative and disruptive business models go even further, allowing companies to sell the function of a product rather than the product itself, and therefore decouple business prosperity from the constant and recurring costs of raw material extraction and production. For example, Rolls-Royce's aerospace division no longer simply sells aerospace engines, but now offers a complete package in which customers purchase the capability that the engines deliver—flight. In this "power-by-the-hour" approach, Rolls-Royce retains the responsibility for risk and maintenance, but simplifies the mechanism for revenue by avoiding the complexities and costs of producing engines in volume, instead coordinating existing stock to be readily available [6].

Herein lies the opportunity: industrial innovators may leverage the natural trend of manufacturing servitization to promote remanufacturing as a sensible strategy for asset protection and economic efficiency. Then, as the servitization trend continues, barriers to the adoption of remanufacturing will naturally be addressed and overcome, expanding its market penetration through voluntary uptake rather than forceful intervention and thus minimizing the cost of transition. In this transition, the expansion of remanufacturing adoption will yield economic, environmental, and social benefits that will enable growing global populations and demand for socioeconomic development to become decoupled from unsustainable resource use.

Figure 5.1 illustrates this relationship between servitization and remanufacturing. In general, the latter faces many barriers to penetration (see (1) in Figure 5.1). Meanwhile, movements toward servitization continue in the conventional manufacturing industry (see (2) in Figure 5.1). This chapter argues that servitization can help reduce such barriers (see (3) in Figure 5.1). In addition, it argues that as the degree of servitization advances, the potential of its effects also increases. As shown in Figure 5.1, the barrier for increasing penetration and benefits (see (3)**) once initial barriers have been overcome is smaller than that inhibiting the benefits at lower levels of uptake (3)*.

This chapter outlines the current state and challenges of remanufacturing and servitization. And explores the incorporation of these two movements. Sections 5.2 to 5.4 outline these

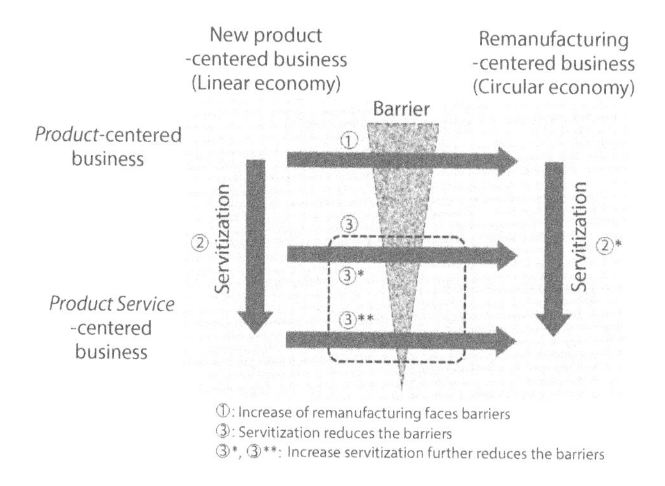

Figure 5.1 Relationship between product service and remanufacturing.

situations and the involved challenges. Section 5.5 presents case studies in two product areas and discusses their implications. Finally, Section 5.6 concludes the chapter.

5.2 Barriers to Remanufacturing

Remanufacturing holds the potential to create numerous benefits for industry and society alike. Among these, it must be considered that remanufacturing is usually more economically feasible than material recycling, as it retains the form and function of the end product and thereby avoids the need to repeat manufacturing processes that create these factors from recycled raw material. This saves labor, equipment, energy, and opportunity costs that would otherwise be spent to produce effectively new products from recycled feedstocks. Remanufacturing also fosters clear material and energy savings compared to new product manufacturing, offsetting not only some of the economic costs of resource consumption, but the environmental impacts as well. Further, because remanufacturing is often a precise undertaking for which few automated processes currently exist, it helps create employment in the manufacturing industry. And finally, since the price of remanufactured products is often lower than that of newly manufactured products, it increases social welfare by providing equitable access to contemporary products and technologies across socioeconomic strata. Based on these benefits, remanufacturing has attracted significant attention from industries, governments, and societies worldwide.

However, remanufacturing has encountered many barriers in its path to global adoption. Much research exists describing the challenges faced by companies in developing a remanufacturing business. Lundmark *et al.* [7], for example, contend that the remanufacturing system is made of three parts—collection, remanufacturing, and redistribution—subject to common barriers:

1. Collecting used products is often difficult and costly
2. Remanufacturing is often complex and costly
3. Customers do not perceive remanufactured products to be of comparable quality or performance to new products, and thus do not accept them

Generally, original equipment manufacturers (OEMs) have advantages over independent remanufacturers (IRs) in resolving these problems due to the control they inherently exert over product designs, production systems, and distribution networks [8]. In contrast, however, OEMs perceive unique obstacles—first in the potential for remanufacturing to compete with, and therefore reduce the sales of new products, and second in the comparatively lower profit margins on remanufactured products that challenge business prosperity [9]. This conflict between new and remanufactured products is referred to as "cannibalization" [10]. This concept is at the center of a fourth major barrier: remanufactured product sales sometimes conflict with new product sales, which discourages OEMs from engaging in remanufacturing (and in some cases even encourages them to actively prevent it), limiting its growth.

To achieve profits in remanufacturing, companies must address these barriers with efforts to circularize processes of resource consumption on their own volition. Moreover, customers, governments, academicians, and societies must support companies in overcoming such barriers by shifting their perceptions of, research in, and engagement with economic norms. Measures in this pursuit include research and development of remanufacturing process technologies (including production management and control systems), the implementation of tools to design products for remanufacturing (DfReman), disrupting marketing and business models with novel approaches, and encouraging circularity through targeted policy initiatives.

5.3 Product Services

In general, products are commoditized in many product areas; as economic globalization shifts jobs across borders to other countries and regions where labor costs are lower, it is becoming increasingly difficult for manufacturing companies in developed countries to win the international cost competition. As a result, companies are seeking ways to achieve higher added value, and product services provide a means to achieve this goal. As stated earlier, many manufacturing companies, especially those based in developed countries, are making efforts to increase the portion of services in their businesses, also known as "servitization" [5]. The factors that promote servitization include the following aspects.

- Services have higher margins than goods [11]
- Services provide a more stable source of revenue than goods due to their resilience against economic cycles [12]

- Services are difficult to imitate [13], and thus companies can lock out competitors [6]
- Provision of service and tailored products develops customer loyalty [6]
- Services fulfill customer needs. In the case of business-to-business trades, customers focus on core competencies and outsource noncore functions, such as maintenance, to providers of capital equipment [14].
- Services are less capital dependent [13].

When manufacturing companies enter the service business, their first step is often to provide repair and maintenance of products. Recently, however, companies have been going further, diminishing the boundaries between manufacturing and service providers. Manufacturing companies are finding that their most important activities are no longer producing and selling as many products as quickly as possible (the traditional model for success in manufacturing), but rather maintaining a core set of existing assets and distributing access to their functionality in an optimized fashion. This requires a significant shift in focus for technology development, workforce expertise, and business management, and is thus a difficult transition for many to navigate.

To this end, it is possible to deconstruct the servitization concept into various classifications of product services in effort to study the most appropriate application context for each. Kotler [15] distinguishes services into two broad categories: maintenance and repair services, and business advisory services. Mathieu [16] also divides services into two groups: services supporting the *product* and services supporting the *customers*. Studies on product-service-systems (PSS) classify product services in a similar manner, with a particular focus on how different levels of product ownership can affect social and economic outcomes. These studies define product services in three broad categories that are particularly useful in thinking about servitization as a strategy to promote remanufacturing: 1) product-oriented services, in which the business model is still largely based on product sales, with some additional services; 2) use-oriented services, in which products remain central but are owned by service providers and made available to users in different forms; and 3) result-oriented services, in which customers and service providers agree on a desired outcome without specifying the product involved [17].

Although the servitization trend seems reasonable from both manufacturers' and customers' perspectives, certain barriers hinder manufacturing companies from effectively implementing servitization. Indeed, empirical research suggests that many manufacturing companies have been unsuccessful in the transfer toward service-based business models [6, 14]. Despite these barriers, customers increasingly expect manufacturers to provide services that, in conventional product-based systems, they would have either done themselves or engaged a third-party to do, creating a patent demand for and economic opportunity in servitization. As global populations continue to place higher value on utility than ownership, the importance of services—and thus the motivation for manufacturing companies to servitize their businesses—will likely increase.

5.4 Product Service as an Enabler of Remanufacturing

This movement towards servitization provides companies with the opportunity to not only increase profits, but also mitigate the barriers of remanufacturing. The possibilities of these effects have been discussed in previous studies, in addition to concepts such as functional economy [18], PSS [17, 19], functional sales [20], service engineering [21], industrial product-service systems (IPSS) [22], and circular economy [23].

The previous section described four types of barriers in remanufacturing. Servitization mitigates these barriers in the following ways:

1. **Collection of used products:** With servitization, returning used products becomes easier. This is especially the case when OEMs retain the ownership of products. Even if other companies (e.g., leasing companies) own the products, reverse supply-chain logistics are simplified when returns come from a few large-scale product owners rather than large numbers of widely-distributed individual end-users.
2. **Remanufacturing process costs:** Servitization enables OEMs to easily monitor the usage of products, enabling them to optimize service intervals based on asset health data. Such optimization minimizes failures, and thus reduces the intensity of required remanufacturing, creating savings in lead-time, labor costs, materials, and energy. When products are sold directly, customers may refuse such monitoring because the transfer of data back to the producer when it is of no immediate valuable use to them is perceived as an unnecessary intrusion on user privacy. In contrast, if OEMs retain ownership, monitoring is relatively easy and often a standardized condition of the use agreement. Even if customers own the products, they are likely to accept usage monitoring if it is associated with valuable services such as preventive maintenance. If the conditions of usage and products are known, remanufacturing is facilitated.
3. **Customer acceptance:** When customers focus on the functions provided by products rather than the products themselves, they become less sensitive to

the appeal of new products. Thus, they are more likely to accept remanufactured products.

4. **OEM motivation:** With servitization, OEMs are encouraged to take responsibility for the entire product lifecycle (rather than just the phases leading up to product sales), because doing so actually helps them reduce costs and improve operational efficiency. This is even more apparent when OEMs retain product ownership; OEMs are motivated to reduce the total cost of ownership to improve profitability for themselves, and remanufacturing provides a means to reduce this cost.

Among these effects, motivating OEMs to remanufacture is significant, as OEMs have clear advantages over independent remanufactures (IRs) in resolving the barriers to remanufacturing [8]. However, OEMs often fear that remanufacturing reduces the sales of new products, causing them to view IRs as competitors and, in some cases, actively work to reduce the possibility of remanufacturing. Similarly, OEMs often perceive that profits from the sales of remanufactured goods are lower than those of new, and are therefore discouraged from remanufacturing themselves. In such cases, OEMs become disablers rather than enablers [24, 25].

Conversely, if OEMs are motivated to remanufacture, they can unlock the vast potential of remanufacturing by implementing DfReman and developing efficient processes, which substantially reduce remanufacturing costs. The prerequisite for this motivation is that remanufacturing costs are lower than new product manufacturing costs.

In addition, remanufacturing provides OEMs with a means to supply customers with products that function the same as newly manufactured products but with lower costs. This indicates not only that servitization facilitates remanufacturing, but also that remanufacturing facilitates servitization. These two processes support one another, and thus, their incorporation constitutes one of the keys to realizing a sustainable, high-value industrial system in our society.

5.5 Industrial Practices

This section describes product services, remanufacturing, and their incorporation in two product areas, using cases from companies primarily based in Japan. These case studies help clarify the prospects and challenges involved in the incorporation of product service and remanufacturing. Product areas of focus include 1) heavy-duty and off-road equipment (HDOR) and 2) photocopiers as a representative of the broader information technology and imaging systems sector. The HDOR case presents an example of the incorporation of product-oriented services and remanufacturing (see Table 5.1, Figure 5.1). The photocopier case illustrates the combination of either use- or result-oriented services with remanufacturing (see Figure 5.1).

5.5.1 Heavy-Duty and Off-Road Equipment (HDOR)

The HDOR equipment remanufacturing sector encompasses companies that remanufacture components of equipment used in construction, mining, farming, and oil and gas drilling industries. Remanufactured components include engines, transmissions, piston pumps and motors, starters, alternators, turbochargers, and hydraulic and electronic system components. The HDOR equipment remanufacturing sector is one of the largest remanufacturing sectors. In the US, the sector's 2011 production value was estimated at 7.8 billion USD, making it the second largest remanufacturing sector behind aerospace [2]. Similarly, in Europe, the sector's production value was estimated at 4.1 billion euro (~4.6 billion USD), making it the third largest remanufacturing sector in the region behind aerospace and automotive parts [3].

Table 5.1 Typology of product services in PSS studies (adapted from [17]).

Type of service	
Product-Oriented	• Product-related services: Provider sells the products as well as the services required during the use phase (e.g., maintenance contract, financing scheme, take-back arrangement). • Advice and consultancy: Provider gives advice on the most efficient use of the products.
Use-Oriented	• Product lease: Provider retains ownership of the products and is often responsible for maintenance/repair. User pays a regular fee, normally for unlimited individual access. • Product renting or sharing: Provider retains ownership of the products and is often responsible for maintenance/repair. User pays a regular fee but does not have unlimited and individual access. The same products are used sequentially by the user. • Product pooling: Provider retains ownership of the products and is often responsible for maintenance/repair. User pays a regular fee but does not have unlimited and individual access. The same products are used simultaneously by the user.
Result-Oriented	• Pay-per-service unit: Product still forms the basis of the PSS. User purchases the output of the products according to the level of use. • Functional result: Provider and user agree on an end result without specifying how the result is delivered.

Caterpillar Inc., based in the US, is the largest OEM of HDOR equipment. Caterpillar created its remanufacturing division in the early 1970s, and it has exhibited sound growth ever since [23]. Komatsu Ltd., based in Japan, has the second largest share in the market, while Hitachi Construction Machinery Co., Ltd. is the second largest OEM in Japan. Komatsu and Hitachi Construction Machinery both conduct remanufacturing in growing capacity; Komatsu's volume of remanufactured components increased fourfold from 2004 to 2017 [26].

The capital costs of HDOR equipment is generally high; construction machines, for example, usually range from tens to hundreds of thousands of USD, while mining machines range from hundreds of thousands to several million USD. Beyond this, the lifecycle costs for after-sales services like repair, maintenance, and consumable parts are usually 1.0–2.5 times the capital costs of the products themselves. As a result, remanufacturing is widely employed to restore many parts and systems rather than replace them, creating significant economic advantage without sacrificing lifecycle performance. Remanufactured components, for example, usually cost 40% less than newly manufactured components, while their lifetimes and warranties are effectively—if not guaranteed to be—identical. This performance equivalence thus creates a relatively steady and predictable market demand for remanufactured components, incentivizing OEMs to engage in remanufacturing because the threat of new product sales cannibalization is reduced. However, strong competition exists between OEMs and third-party companies who aim to offer refurbished components at even lower prices.

Among the aforementioned barriers, collection logistics and customer acceptance are typically the most challenging. However, in this case, challenges in customer acceptance are not caused by competition between remanufactured products and new products, as is the case in most other markets, but rather by competition between OEM- and third-party-remanufactured products that creates customer hesitation based on the uncertain balance between cost and compatibility. This competition also complicates collection logistics, since used product recovery is most commonly performed by third-party companies who must balance the economics of supplying product cores either to a network of multiple smaller, independent remanufacturers or directly to a more concentrated pool of large OEMs.

Meanwhile, OEMs have been aiming to increase their revenues from services. Although the potential revenue available from after-sales service provisions is large, they have not sufficiently realized this potential, partly because competitors (i.e., third-party service providers) often provide after-sales services, including spare components, as a much more central component of their business model. In response, some OEMs have developed to remote asset health monitoring systems in effort to support servitization. These systems monitor the position and operational status of deployed machines via telecommunications technology, allowing them to identify eminent failures or optimal remanufacturing. These systems enable the OEMs and contracted after-sales service providers to increase the efficiency services by automating the determination of when, where, and why specific supplies are needed. Knowledge of machine status also helps extend product lifetimes by informing service providers of emerging maintenance needs before failures occur, thus mitigating overall wear on the product. This preventative maintenance model also enables better production planning and inventory control for remanufacturing. This, in turn, generates a competitive advantage over third-party service providers, reduces service provision costs, increases customer satisfaction, and reduces environmental load. Currently, OEMs are exploring new areas of services in effort to expand businesses even further.

Currently services in the HDOR industry are categorized as product-oriented (see Table 5.1). In this case, strengthening services has enabled OEMs to increase the sales of spare components in the aftermarket as an independent revenue stream, and it has simultaneously increased the supply of remanufactured components in the market. In addition, OEMs are continuously exploring new areas of services. As a result, the potential use of remanufactured components as well as OEM commitment to product life-cycle management will likely increase.

5.5.2 Photocopiers

The photocopier industry has been a forerunner both in servitization and remanufacturing. Since the inception of the industry in the 1950s, OEMs have constructed a business model that combines products and services. Revenue in the photocopier business is generated from three areas: 1) sales of products; 2) sales of supplies; and 3) repair and maintenance services. The most mature, and perhaps still most prevalent revenue model is to sell the product to a customer (i.e. transfer ownership) and continue revenue generation by subsequently selling maintenance services and replacement consumable supplies (such as toners, drums, and parts) that must frequently be replaced. To this end, usage-based service agreements are often used in which service intervals and rates are defined according to a set number of printed pages. More recently, however, leasing models have emerged in which the OEM retains ownership of the machine, and customers simply pay for its use over a defined period. This reduces the capital costs incurred by customers, as well as the material, energy, and process costs of continuously manufacturing new products incurred by the OEM. In either case, these services are categorized as use-oriented, and have a demonstrated record of both economic viability and environmental preferability in many cases. The history of servitization in this industry has been studied and cited extensively as a prime example of servitization [5, 27].

To support these business models, photocopier OEMs have been conducting remanufacturing for more than a decade. In Japan, Fuji Xerox Co., Ltd. started remanufacturing in the 1990s, followed by industry competitors Ricoh and Canon Inc.

in the 2000s. These companies conduct two types of remanufacturing, each with distinct characteristics [25]. Ricoh and Canon, for example, refurbish components from used products to make a wholly remanufactured machine. This expands the potential market base to such an extent that profits from remanufactured machines are in many cases larger than those from virgin machines, making the business model viable despite some degree of competition. In contrast, Fuji Xerox incorporates remanufactured components into new products. In this case, virtually all products may include remanufactured components, thus abolishing the distinction between new and remanufactured products and eliminating any intra-company competition. In both cases, the companies implement DfReman strategies as a deliberate means to facilitate their own remanufacturing operations, substantially enhancing the efficiency of their processes. Over many years of deployment, the expertise gained in this area has allowed remanufacturing to become the norm in business practice, and remanufactured products to become the norm in the user market.

In this industry where remanufacturing is already well-established, the introduction of product services, especially in the form of a PSS-leasing model, show great potential to facilitate the collection of used products for OEMs, and thus make remanufacturing even simpler and more profitable. In Japan, OEMs take back more than 80-percent of products sold to customers after the products are used. In addition, the amount of remanufactured products supplied by OEMs is increasing in Japan, as well as in other developed countries. Ricoh, for example, developed a new remanufacturing facility in Japan in 2016 to increase the capacity of remanufacturing to 20,000 photocopiers per year (Figure 5.2).

Meanwhile, OEMs' incentives to increase remanufacturing still have room to expand. One reason for this is that the comparatively lower price of remanufactured products encourages more customers to buy them, aligning well with OEM objectives to increase revenues. Another reason is the complex cost

Figure 5.2 Ricoh's remanufacturing facility (Gotenba, Japan).

relationships inherent in remanufacturing and reusing parts. While the profit margins of remanufactured products are in many cases as high as that of newly manufactured products, this parity only continues up to a certain number of products. As the volume of remanufactured products increases, it becomes increasingly difficult to identify and collect a supply of used products in sufficient quality, consequently increasing the cost of remanufacturing. As the cost increases, the profit margin on remanufactured products narrows, thereby incentivizing the OEM to rely instead on new product sales. Further advancement of servitization may change this situation by making sales-volume-based business models all but incompatible with both customer demands and use-oriented value propositions, and thus rendering new production economically unviable.

To this end, services in this industry have evolved in recent decades. A service called managed print service (MPS), for example, appeared in the industry in the late 1990s first from Xerox (US) and Fuji Xerox (Japan), and then by both Ricoh and Canon. In general, MPS refers to solutions that aim to optimize and manage customer document output environments by merging the lease of hardware and the provision of service into a single unified contract system. Under a traditional contract, a customer pays for products and services separately; in contrast, MPS charges a customer for product lease and services in a singularly-defined scheme. For example, rather than customers purchasing or leasing products *and* paying for services according to use intensity (e.g., 1 cent per copy), an MPS contract charge covers both hardware lease and service costs by merging them into a single, higher fee and changing its nominal definition. Thus, the effective price per copy might be higher than that in the former case (e.g., 3 cents per copy), but the customer is relieved of responsibility for capital hardware costs because the additional per-copy price covers the equipment cost. In both cases, OEMs typically guarantee the functional performance as a measure of value (e.g., downtime less than 5%). In this, MPS is categorized as a result-oriented service (Table 5.1), similar to Rolls-Royce's "power-by-the-hour" business model. In this type of contract, OEMs have incentive to minimize product lifecycle costs because doing so directly corresponds to an increase in overall extractable value, so long as products can achieve the guaranteed level of performance. This has led OEMs to use remanufactured products and even refurbished products as key components of the service business model because they can provide new-equivalent performance as significantly reduced lifecycle cost to the OEM. With regard to the latter, refurbishing is similar to product reuse in that restorative interventions are minimal, and product quality is not guaranteed to be as good as new. However, the cost for refurbishing is understandably lower than that for remanufacturing, though there is greater performance risk. OEMs sometimes assume the risk that used products will perform adequately even without complete remanufacturing, and thus refurbish used photocopiers for use in MPS plans to continue extracting value while lowering interim process costs to themselves (Figure 5.3). In such result-oriented services, OEMs derive revenue exclusively from product function; they must therefore assume responsibility for (and aim to minimize) complete lifecycle costs, in turn, motivating them to remanufacture and refurbish.

Figure 5.3 Photocopier refurbishing in Canon Singapore.

The collection of used products is a well-addressed barrier in the Japanese market; however, it is still a considerable challenge in other countries. Even in developed countries in Europe and North America, OEMs cannot collect approximately half of used products economically. As a result, many of these products go to developing countries—sold en masse for a small portion of their potential remaining value—where they are refurbished by third-parties (Figure 5.4). Moreover, OEMs in developing nations cannot collect most of the used products, as the barrier of remanufacturing process costs has also been a challenge. Although the profit of remanufactured products on part with that of new products, the expected benefits are long-term, preventing small companies in developing nations from finding enough immediate economic stability to engage in remanufacturing. Servitization, of course, exhibits great potential to mitigate all of these barriers by shifting the focus of profitability away from product volume (i.e. long-term benefits) and towards service provision with inherently immediate economic returns. While this case exemplifies

Figure 5.4 Photocopier refurbishing by a third-party company in Indonesia, a developing country.

the effects of servitization in promoting remanufacturing, there is still much room for improvement in the industry.

5.5.3 Summary and Implications

In recent decades, HDOR OEMs have made considerable efforts to strengthen after-sales services to support sales-based business models. As the provision of spare parts in the aftermarket increases, the provision of remanufactured parts also increases. In contrast, photocopier OEMs have provided after-sales services since the inception of the industry, and the products are usually provided through leasing. In both cases, the capital-intensive nature of the products and the function-focused nature of the users have encouraged OEMs to collect used products and start remanufacturing as a means to both cut costs and fulfill customer demands in an expanding market base. New business models centered upon service delivery as the primary mechanism for generating revenue exemplify result-oriented services and provide further opportunities for OEMs to promote remanufacturing and refurbishing.

While servitization helps mitigate the aforementioned barriers to growth in remanufacturing, the difficulty capturing the benefits of this mutually-dependent system is largely determined by product properties, especially design, functional complexity, and the scale of product markets that affect the intensity (and thus cost) of reverse logistics and remanufacturing. Although remanufacturing reduces costs by reusing subcomponents and avoiding the material, energy, and process costs of virgin production, it remains a skill- and labor-intensive process with uncertain supply chains, in contrast to automated mass production for new products and well-established forward distribution networks. Cost advantages and disadvantages are in large part determined by these balances. In the aforementioned cases, the cost advantage of remanufacturing is most likely higher in the HDOR sector than in photocopiers. Figure 5.5 illustrates the servitization level and cost advantages of remanufacturing in several product areas. The aerospace industry, for example, is advanced in servitization and the cost advantage of remanufacturing is also high due to the similarly high capital costs of new products. Conversely, while servitization in the auto parts industry is similar to that in the HDOR industry, the cost advantage of remanufacturing is less apparent due to lower capital costs for new replacements relative to the

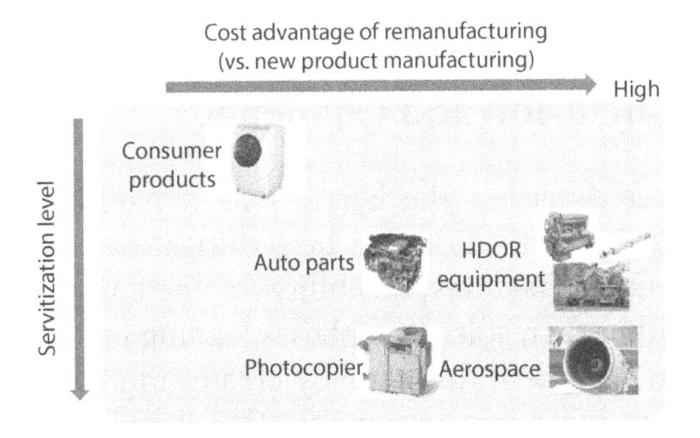

Figure 5.5 Mapping the servitization level and cost advantage of remanufacturing.

HDOR or aerospace industries. As a result, auto parts OEMs are comparatively less active in remanufacturing, and third-party remanufacturers tend to dominate the market in many countries. Consumer products, in contrast, are considerably more challenging both in servitization and the cost effectiveness of remanufacturing.

While the difficulties of remanufacturing caused by intrinsic product properties should be taken for granted, the manufacturing industry's movement toward servitization is an ongoing trend. In addition, the wave of the Internet of Things (IoT) in recent years offers an opportunity for the industry to unlock the potential of servitization on an unprecedented scale [28]. Even in consumer product areas, consumers seek more to pay for product functions than for ownership of the products themselves. The globally growing market for services such as car sharing, rentals, leases, and carpooling in the automotive sector, for example, exemplify this and should be a clear signal to OEMs that new opportunity in PSS business models supported by remanufacturing awaits [23]. This leveraging of servitization to capture new markets and simultaneously expand the applicability of remanufacturing can be extrapolated across quite nearly any manufacturing industry sector. In sum, the trend of servitization can help mitigate the barriers of remanufacturing, and it therefore demands further exploration and development.

5.6 Conclusion and Challenge

This chapter describes the barriers to remanufacturing and explores how servitization can help mitigate them. With servitization, OEMs take responsibility for the complete product lifecycle rather than solely the phases leading up to its sale; but rather than a burden, this actually creates mutual benefits for both the producer and consumer. This is even more apparent when OEMs retain product ownership in use-oriented or result-oriented product-service-systems. In these models,

OEMs are motivated to reduce the total costs of product ownership in effort to maximize the value they can extract from its life. To this end, remanufacturing and refurbishing provide a viable means to reduce these costs. The spread of servitization, then, can be effectively used as a vehicle to advance the global uptake of remanufacturing in pursuit of a more sustainable industrial economy.

For servitization to become the driving force for remanufacturing, however, the cost-effectiveness of remanufacturing must become better than that of new product manufacturing. Although remanufacturing is potentially more cost-effective than new product manufacturing since it avoids the costs of new material, process energy, and advanced manufacturing technology, its cost-effectiveness is not universally apparent due largely to complexities and uncertainties in reverse logistics and closed-loop EOL product supply chains. In addition, cost effectiveness differs depending on product design properties, functional technologies, and other related factors. Thus, it is important for interested parties—i.e., manufacturing companies, remanufacturers, academicians, governments, and societies—to make concerted efforts in further increasing the cost effectiveness of remanufacturing in the future.

Despite the aforementioned advantages, various challenges still exist. First, developing methodologies and implementation strategies for DfReman is crucial to enhancing the cost effectiveness of remanufacturing. Second, developing remanufacturing processes techniques to meet 21st century technologies and workforce skillsets is necessary to maintaining competitiveness in an increasingly fast-paced and complex production environment. Such techniques include those for product disassembly, failure assessment, subcomponent cleaning, material surface treatment (including additive manufacturing), performance testing, and remote monitoring and diagnosis. Third, developing techniques to address the uncertainty quality, quantity, and timing of returned products is essential to establishing effective schema of operational management in remanufacturing processes that are equipped to contend with often more predictable and simplified systems in conventional manufacturing. Fourth, policies to support and, indeed, encourage the adoption remanufacturing across industry sectors are required. This includes trade regulations for used products, local market influence, and extending producer responsibility through environmental legislation. Finally, developing rules and systems that motivate both OEMs and third-party remanufacturers to promote remanufacturing is necessary.

In all, the potential for remanufacturing to create significant economic, environmental, and social benefits is immense, and its global adoption be vital to our transition to a more circular economy. Based on these prevalent issues, however, it is essential that our generation realize the high value of industrial servitization and remanufacturing both independently, and as cooperative elements in a larger movement towards global economic and environmental efficiency and sustainability.

References

1. Matsumoto, M. and Nasr, N., Remanufacturing as an enabler for green service models, in: *Service and the Green Economy*, A. Jones et al., (Ed.), pp. 75–98, Palgrave Macmillan, London, 2016.
2. US International Trade Commission (USITC), *Remanufactured goods: An overview of the U.S. and global industries, markets, and trade*, Investigation No. 332-525, USITC Publication 4356, http://www.usitc.gov/publications/332/pub4356.pdf, 2012.
3. European Remanufacturing Network (ERN), *Remanufacturing market study*, https://www.remanufacturing.eu/wp-content/uploads/2016/01/study.pdf, 2016.
4. Matsumoto, M. and Yang, S., Martinsen, K., Kainuma, Y., Trends and research challenges in remanufacturing. *Int. J. Precis. Eng. Manufact.-Green Technol.*, 3, 1, 129–142, 2016.
5. Vandermerwe, S. and Rada, J., Servitization of business: Adding value by adding services. *Eur. Manage. J.*, 6, 4, 314–324, 1988.

6. Neely, A., Exploring the financial consequences of the servitization of manufacturing. *Oper. Manage. Res.*, 1, 2, 103–118, 2008.
7. Lundmark, P., Sundin, E., Bjorkman, M., Industrial challenges within the remanufacturing system. *Proc. Swed. Prod. Symp.*, 132–139, Stockholm, 2009.
8. Lund, R. and Skeels, F., *Guidelines for an original equipment manufacturer starting a remanufacturing operation*, Massachusetts Institute of Technology, Center for Policy Alternatives, Cambridge, MA, 1983, Government Report, DOE/CS/40192, CPA-83.8.
9. Linton, J., Assessing the economic rationality of remanufacturing products. *J. Prod. Innov. Manag.*, 25, 3, 287–302, 2008.
10. Guide, V.D.R. and Li, J., The potential for cannibalization of new products sales by remanufactured products. *Decis. Sci.*, 41, 547–572, 2010.
11. Anderson, E.W., Fornell, C., Rust, R.T., Customer satisfaction, productivity, and profitability: Differences between goods and services. *Mark. Sci.*, 16, 2, 129–145, 1997.
12. Quinn, J.B., Doorley, T.L., Paquette, P.C., Beyond products: Services-based strategy. *Harv. Bus. Rev.*, 68, 2, 58–67, 1990.
13. Wise, R. and Baumgartner, P., Go downstream: The new profit imperative in manufacturing. *Harv. Bus. Rev.*, 77, 5, 133–141, 1999.
14. Oliva, R. and Kallenberg, R., Managing the transition from products to services. *Int. J. Serv. Ind. Manag.*, 14, 160–172, 2003.
15. Kotler, P., *Marketing management: Analysis, planning, implementation, and control*, Prentice Hall, New Jersey, 1997.
16. Mathieu, V., Product services: From a service supporting the product to a service supporting the client. *J. Bus. Ind. Mark.*, 16, 39–61, 2001.
17. Tukker, A., Eight types of product-service systems: Eight ways to sustainability? Experiences from SusProNet, in: *Business Strategy and the Environment 13*, pp. 246–260, Wiley, Hoboken, NJ, 2004.
18. Giarini, O. and Stahel, W.R., *The Limits to Certainty - Facing Risks in the New Service Economy*, Kluwer Academic, Boston, 1989.
19. Goedkoop, M., van Halen, C., te Riele, H., Rommens, P., *Product service systems, ecological and economic basis*, Pricewaterhouse Coopers N.V./PiMC, Storrm C.S., Pre consultants, Amersfoort, Netherlands, 1999.

20. Sundin, E. and Bras, B., Making functional sales environmentally and economically beneficial through product remanufacturing. *J. Clean. Prod.*, 13, 9, 913–925, 2005.
21. Sakao, T. and Shimomura, Y., Service engineering: A novel engineering discipline for producers to increase value combining service and product. *J. Clean. Prod.*, 15, 6, 590–604, 2007.
22. Meier, H., Roy, R., Seliger, G., Industrial product-service systems - IPS2. *Cirp. Ann.-Manuf. Techn.*, 59, 607–627, 2010.
23. Ellen MacArthur Foundation, *Towards the circular economy*, https://www.ellenmacarthurfoundation.org/assets/downloads/publications/Ellen-MacArthur-Foundation-Towards-the-Circular-Economy-vol.1.pdf, 2013.
24. Seitz, M.A., A critical assessment of motives for product recovery: The case of engine remanufacturing. *J. Clean. Prod.*, 15, 11&12, 1147–1157, 2007.
25. Matsumoto, M. and Umeda, Y., An analysis of remanufacturing practices in Japan. *J. Remanufacturing*, 1, 2, 1–11, 2011.
26. Komatsu, Promoting the Reman Business, http://home.komatsu/en/csr/environment/recycle/reman.html (Accessed on August 8, 2019), 2018.
27. Visintin, F., Photocopier industry: At the forefront of servitization, in: *Servitization in Industry*, G. Lay (Ed.), pp. 23–43, Springer, New York, 2014.
28. Rymaszewska, A., Helo, P., Gunasekaran, A., IoT powered servitization of manufacturing - an exploratory case study. *Int. J. Prod. Econ.*, 192, 92–105, 2017.

6

Design for Remanufacturing

Brian Hilton*[*] and Michael Thurston[†]

Golisano Institute for Sustainability, Rochester Institute of Technology, Rochester, NY

Abstract

A viable industrial strategy for decades, remanufacturing is reemerging as a key enabler of the modern transition to a circular economy. In order to encourage widespread adoption of remanufacturing, however, it is vital to consider not only the business models, supply chains, and market systems that support it, but also, fundamentally, what makes products remanufacturable. In this chapter, we explore the concept of Design for Remanufacturing (DfReman), its role in both new product development and remanufacturing business models, and the major principles that constitute the framework for an effective Design for Remanufacturing strategy. We suggest that there are three such principles: designing to (1) create value, (2) protect and preserve value, and (3) easily and cost effectively recover value. Each of these principles incorporates multiple supporting concepts, which are discussed in the latter sections of this chapter. Ultimately, we posit that design strategy is at the center of the remanufacturing equation. In other words, the economic, environmental, and technical feasibility of building a circular economy through remanufacturing effectively hinges on products being designed specifically to be remanufactured.

[*]*Corresponding author*: bshasp@rit.edu
[†]*Corresponding author*: mgtasp@rit.edu

Nabil Nasr (ed.) Remanufacturing in the Circular Economy (137–168)

Keywords: Remanufacturing, design for remanufacturing, barriers, enablers, end-of-life

6.1 Introduction

Society and industry now operate in a resource constrained world, driven in part by increasing demand for natural resources in developing nations and continued strong demand in the developed world. In just the last fifty (50) years, the Earth's population has more than doubled, and presently, nearly 7.5 billion people compete for the Earth's dwindling supply of fossil fuels, clean fresh water and material resources [1]. In just the last five (5) years, the population has increased by over 400 million people [1]; greater than the entire population of the United States [2]. As the population grows, there is further competition for the use of earth's natural resources and demand on its ecological systems. And it is not just the increase in world population that is putting a strain on the environment, but also the improvement in people's standard of living. The GDP of China for example has increased 422 percent between 1990 and 2008 [3].

There is a great social benefit to any population raising themselves out of poverty; however, given current practice, an increased standard of living comes with an increased global impact. For example, the United Nations Environment Programme has noted that a doubling of income is related to 81% higher CO_2 emissions [4]. Greater access to modern technologies and the related increase in consumption is a component in raising the standard of living of less developed societies. At the same time, driven by technological change the use cycles of many products are shrinking. These factors are resulting in increased material demands and environmental impacts. They are beginning to generate new constraints on the designs of future products as governments, academia, and industry look for ways to decouple development from environmental impacts.

As businesses and governments try to address the problem of diminishing resources, there is now a considerable interest in sustaining and expanding the remanufacturing industry as a means of reducing the material consumption and other impacts associated with modern products. Remanufacturing (reman) is an industrial process that restores worn and discarded products or modules to a like-new condition. The restoration is a high-quality industrial process through which products are systematically disassembled, cleaned, and inspected for wear. Damaged components are replaced or restored, feature upgrades can be incorporated, and the product is reassembled. Finally, product testing ensures that performance will meet specifications [5].

Remanufacturing is able to recover and preserve much of the raw materials and the value added processes (the labor, energy, and manufacturing processes) that were embodied in the original product. This is often both environmentally and economically beneficial. Typically at the end of its life, a product would be destined for landfill, incineration or recycling. But by implementing a remanufacturing strategy, disposal costs can be avoided, the value still embodied in a product can be recouped, and natural resources can be more efficiently utilized. Incineration only recovers the energy value embodied in a material (for example, plastics), while recycling recovers the raw material value. As noted above, remanufacturing can recover most of the energy and economic value embodied in the product manufacturing process.

It is widely accepted that the United States has one of the most diverse and developed remanufacturing industry sectors in the world [6]. Much of this development stems from the industrial revolution and the subsequent invention of mass production. A good example can be seen in the history of Ford Motor Company. Once Ford had produced and sold millions of Model-T cars in the early 1900's, it made practical business sense to create separate facilities to remanufacture failed motors, and a distribution system to support Ford customers with economical like-new engines and service parts [7].

This pragmatic approach to integrating remanufacturing into global product support strategies continues to this day. Thousands of product categories including office furniture and equipment; transportation; construction and electrical equipment; medical devices; machine tools, compressors, heavy machinery, and others, account for the global remanufacturing business activity. Major corporations such as Caterpillar, Cummins, Flextronics, Xerox and General Electric all generate significant business through remanufacturing programs related to their products.

Though the remanufacturing industry was pioneered over 80 years ago in the United States, it is still enjoying significant growth. The U.S. is the largest remanufacturer in the world, and in 2011, the value of U.S. remanufactured production grew to over $43.0 billion and supported 180,000 full-time U.S. jobs. U.S. exports of remanufactured goods also totaled $11.7 billion in 2011 [6]. The reman industry is diverse, with the twelve top remanufacturing-intensive sectors that account for the majority of remanufacturing activity in the United States [6] shown in Figure 6.1.

Additionally, the U.S. remanufacturing industry's largest sector is the defense industry with a significant interest in sustaining a substantial number of ground vehicles, ships, aircrafts, and other support systems.

As society transitions to greater sustainability in the decades ahead, remanufacturing will continue to be a powerful driver

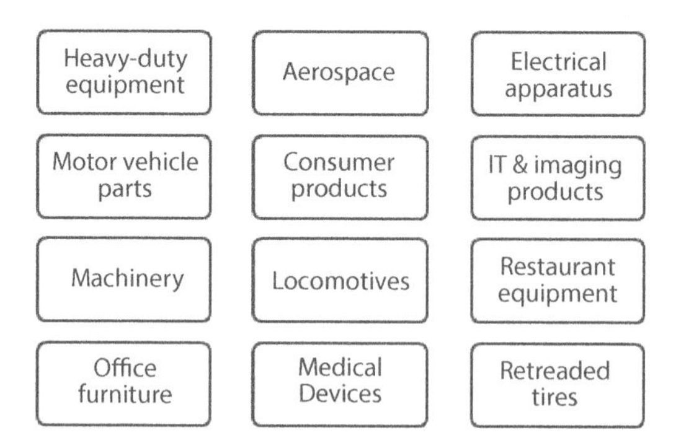

Figure 6.1 Twelve remanufacturing sectors.

for improvement. However, to capitalize, industry must address many of the barriers it faces that are restricting growth of remanufacturing.

6.2 Defining the Barriers to Remanufacturing Growth

The Golisano Institute for Sustainability (GIS) at the Rochester Institute of Technology (RIT) was contracted by the National Institute of Standards and Technology (NIST) at the U.S. Department of Commerce in support of the NIST Advanced Manufacturing Technology Consortia (AMTech) Program Opportunity 2014-NIST-AMTECH-01. The goal of this program was to develop a roadmap and consortium framework to support long-term, pre-competitive research relevant to the remanufacturing industry. A part of the program deliverables was to provide an actionable plan to overcome technological barriers currently inhibiting the growth of the industry. The program sought to promote advanced manufacturing capabilities in U.S. remanufacturing in light of growing constraints related to resource availability, manufacturing waste, and energy use on a global scale. Because the remanufacturing industry is highly diverse—spanning 12 sectors and including varied organizational structures (e.g. original equipment manufacturers-OEMs and independent remanufacturers, both large companies and small and medium-sized enterprises (SMEs)—it was important to establish a baseline level of mutual understanding of both the goals of the study and of industry barriers and opportunities.

GIS therefore brought together an Advisory Group to help define potential barriers to growth and focus areas in which pre-competitive collaboration may create industry-wide benefits. These focus areas included design and process technology, reverse logistics (e.g. tracking and identification, transportation, acquisition, and valuation), market awareness & recognition, legal and legislative issues, workforce education & training,

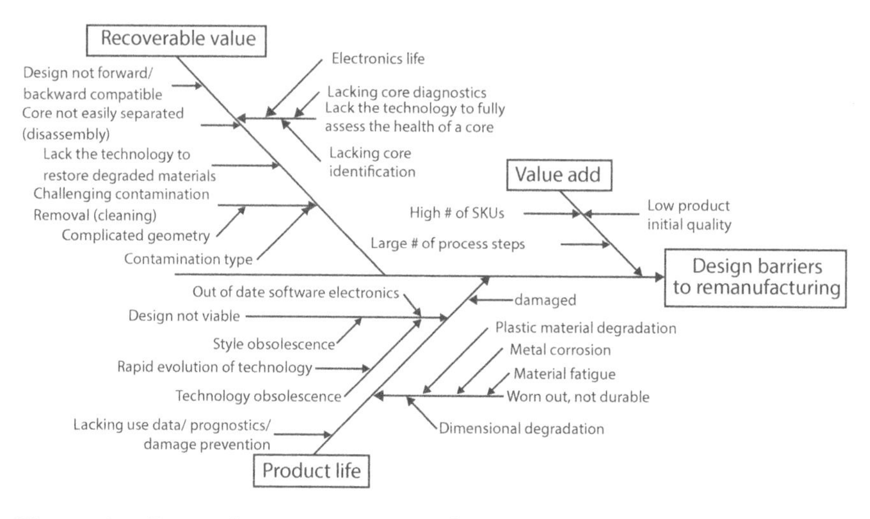

Figure 6.2 Design barriers to remanufacturing.

market access & finished goods, and policy integration. The Advisory Group was comprised of leaders from prominent remanufacturing industry groups, leading remanufacturing industry corporations, remanufacturing experts and researchers from academia and national labs.

The GIS team used the information discussed during the Advisory Group meeting to develop a cause-and-effect diagram (or fishbone diagram) capturing the causes and barriers to remanufacturing growth. The Group identified barrier topics ranging from workforce issues to market restrictions, management support to reverse logistic challenges, operational integration to lack of design features. Topics also included market awareness and recognition, education and training, and legal and legislative issues. The issues that were highlighted specifically related to product design are summarized in a fishbone diagram (Figure 6.2).

6.3 Remanufacturing Design Enablers

The design barriers identified by the Advisory Group align well with a strategic approach to remanufacturing that was proposed

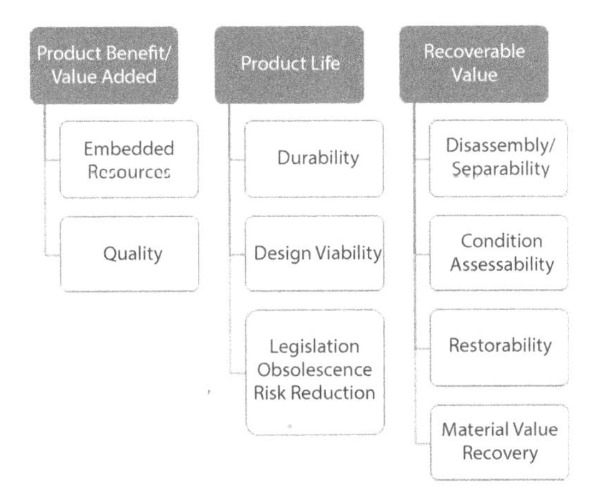

Figure 6.3 Remanufacturing design enablers.

by the authors in a prior article [5], in which key enablers to a remanufacturable design are identified. The framework for this approach included design enablers and business enablers and this framework is recreated here in Figure 6.3 with only the design enablers featured.

6.4 Three Principles of Designing for Remanufacturing

When merging the barriers provided by the Advisory Group and the enablers from the prior article, three major principles to designing for remanufacturing materialize: (1) design to create value, (2) design to protect and preserve value, and (3) design to easily and cost-effectively recover value. This is illustrated in Figure 6.4.

Before applying these design principles, the product should be judged as suitable for remanufacturing. As noted in Nasr *et al.* [5], at its heart, remanufacturing is a strategic decision that is not solely based on the design of a single product. For example, a company must decide whether it has the know-how and the infrastructure to support remanufacturing. The product design must also be viable far enough into the future to achieve the

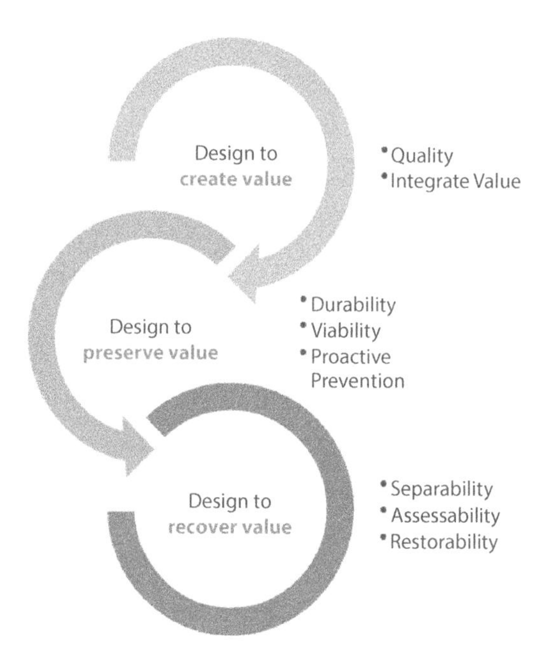

Figure 6.4 Design for remanufacturing three principles.

desired contribution of product recovery and remanufacturing to the company's business case and sustainable production metrics.

The remainder of this chapter discusses specific design for remanufacturing principles with the assumption that any product under consideration has already passed high level technical and business feasibility criteria. There has been significant prior research in the area of design for remanufacturing, and the rest of this chapter will expand upon the three defined design principles in light of prior research.

6.4.1 Design to Create Value

The first design for remanufacturing principle is to create value that is worth recovering. Value is created through the resources and effort required to invent and refine product concepts that provide desired or needed functionality, combined with the materials, processes and tooling required to produce the product efficiently.

Products and components have embedded investments in development effort, and in tooling, assembly time, and materials. Products are removed from service after they have physically failed, or when they no longer meet the functional demands of the marketplace (features, performance, reliability, etc). Product life extension through remanufacturing recovers and re-utilizes the embedded manufacturing value and resources. Products that create high market value and market demand by meeting or exceeding performance, quality, and reliability expectations will be better candidates for remanufacturing.

Products that are designed with a robust design philosophy deliver consistent performance in the presence of component manufacturing variation and will therefore also be easier to remanufacture. Another key factor in creating value is how a particular product design fits into an overall product strategy which may include families of like products. The following subsections highlight design guidelines used to create value in a product.

6.4.1.1 *Designing for Product Quality*

Designing a product to be remanufactured starts with setting high standards for the quality and function of the initial product. Understanding functionality at the product and system levels and how it is affected by feature and manufacturing variation is at the core of Quality Function Deployment (QFD). QFD, originally developed by Akao [8], provides a framework for prioritizing and relating customer requirements and technical performance descriptors, and balancing these requirements when they may conflict. Following a structured framework such as QFD can help to insure that the functional value is embedded in the original product. How these requirements may change over time, due to customer expectations or market conditions, should also be considered. Yüksel describes a QFD House of Quality for remanufacturing, however the "voice of the remanufacturer" requirements that are described [9] are more applicable to the categories of preserving and recovering value.

Concurrent development of product designs and manufacturing processes has demonstrated improved quality and reduced cost of manufactured products [10]. One concurrent engineering process that has shown results, particularly with respect to product quality is error proofing, also known as Poka-yoke [11]. By understanding how manufacturing processes can fail, design countermeasures can be taken to prevent these failures from occurring. Another process that can be implemented to improve quality through design is robust design [12]. In defining design feature values, or manufacturing tolerances, care should be taken to understand the impact of variation on critical to quality product features. Tolerances also need to understand manufacturing process capabilities.

Some key product design guidelines, with respect to insuring lasting product quality, are described below.

- *Understand product requirements, and how they may change over time, especially critical to quality performance factors.*
- *Simplify the design to use the fewest parts and processing steps. Simple designs result in inherently higher quality products.*
- *Understand quality problems in previous products to prevent new product development teams from repeating past mistakes.*
- *Mistake-proof the design and assembly.*
- *Use Multi-functional design teams to ensure that all quality issues are raised and resolved early in the design process.*
- *Optimize tolerances in a systematic way to achieve high quality at a low cost. Identify critical dimensions that need tight tolerances and precision parts, and loosen tolerances on cheaper parts.*

- *Reuse proven designs and processes to minimize risk, maximize reuse, and assure quality.*
- *Standardize and use common parts.*

6.4.1.2 Integrate Value

Another design path to create product value is by integrating functions into high value components and systems that are accessible and easy to recover. This consolidation reduces the overall number of parts, reduces the assembly and disassembly complexity, and also improves serviceability.

Salhieh and Kamrani [13] describe a structured process of design for modularity; a process that considers how function and physical characteristics should be decomposed and clustered. They also describe how modularity enables product platforms; products built around a particular technological platform. Modular design encourages standardization, and also is an enabler for families of products developed around a particular platform [14].

As described by King and Burgess [15], platform design and commonality are enablers of remanufacturing. Standardization across product platforms or product families, particularly for high value components, provides a larger potential source of remanufacturing raw materials (core components).

Specific design for remanufacturing guidelines to integrate product value include:

- *Design modular products to facilitate component standardization and ease of assembly and service.*
- *Reduce the number of components by designing components to be multifunctional, incorporating multiple functions. Fewer components make disassembly easier and faster.*
- *Multi-product compatibility/commonality is essential to maximize the utilization of modules and components. This entails their use in different configurations*

such as new products of the same model, or different models within the product family.

6.4.2 Design to Preserve Value

The next design for remanufacturing principle is to design to protect and preserve the product's value throughout its use so that the function can be restored for use in additional life cycles. This principle includes designing in durability so that the product resists material degradation, corrosion, and wear, and provides protection against damage during use and recovery.

Preserving value also requires that the design is viable at the time of recovery, either through timeless design or through upgradeability of the product's performance, aesthetics, technology, or materials. The viability of product design elements may be impacted by changes in customer requirements, market competition, or changing regulation.

Finally, the design should preserve the product's value by monitoring performance during use, and preventing catastrophic failure that would render the product worthless upon recovery. The following subsections highlight guidelines used to preserve value.

6.4.2.1 Designing for Durability

Designing to preserve the value during use starts with building durability into the product. Many products fail during use due to material degradation including corrosion, wear, abrasion, and fatigue. Additionally, products receive damage not only during use, but also during recovery and processing.

With respect to damage incurred during use, parts can be made more resistant to fatigue by material selection or adding additional material. They can be made more resistant to wear by selecting different materials or coatings. Design is a process of balancing trade-offs between often conflicting requirements.

Requirements around weight and product cost may out-weigh the benefits of increased durability. Bras and Hammond [16] suggest isolating wear to parts that have less value, and are therefore more economical to replace.

The key design activity is evaluating the component durability against the anticipated use-cycle of new and remanufactured products. Cost trade-offs should be made considering the extended product life cycle, including potential for multiple product/component remanufacturing cycles. This trade-off should also consider the options for restoration or repair of damage. If appropriate restoration technologies do not exist then it is more important to consider making the component more durable to allow for multiple use-cycles.

Design for remanufacturing guidelines to assure product durability include:

- *Consider design options to extend component life to multiple use cycles (such as alternative materials or processes) and use life cycle cost when evaluating trade-offs.*
- *Use corrosion or wear resistant coatings to minimize component damage.*
- *Separate features that are exposed to high wear or fatigue loads from longer life features. Consider the use of sacrificial parts as wear surfaces to protect important components.*
- *Target infinite fatigue life when practical.*
- *Design large components with replaceable wear surfaces rather than permanently cast, molded, or machined-in features to minimize the size of the components to be restored or replaced.*
- *Avoid materials that degrade with age or exposure to environmental conditions such as UV degradable plastics.*
- *Avoid materials that degrade when exposed to known chemicals, such as foam rubber seals exposed*

to cleaners. If these materials are required, make replacement easy.

- *Housings and external features should be designed to minimize damage to product during recovery operations.*

6.4.2.2 Designing for Viability

Many products have requirements that may change over their life cycle, such as functional performance, aesthetics, energy efficiency, or material use restrictions. Products also may contain technology that experiences rapid market change, such as computing hardware and software. Preserving the product's value for multiple life cycles may therefore require that the product be designed for viability at the time of recovery, either through timeless design or through upgradeability. To the extent that these changes can be planned for during the design process, modularity can facilitate technology and performance upgrades.

Computer servers are an excellent example of this design principle. A survey of the IT market by IDC Research revealed that replacing a server after three years of operation will have a return on investment (ROI) of less than one year based on the efficiency, reliability, and performance gains of the new equipment, as compared to continual operation of current equipment [17]. Additionally, the cost for power and cooling grew eight times as fast as server purchasing costs and maintenance and management costs grew four times as fast as server purchasing costs, enhancing the effects of the efficiency gains [17].

Rapid changes in performance like this represent an opportunity for manufacturers to design in features to enable upgrades. Though not commonly remanufactured, server manufacturers have taken this principle to heart and have designed many of the components with improving performance to be "refreshed" or upgraded such as: memory, mass storage devices, network connectivity, processors, and power supplies, allowing for the service life of the product to be extended. This principle can easily be applied to remanufacturing.

Other products, such as the PuzzlePhone (http://www.puzzlephone.com/), have been designed with various technology subsystems concentrated in modules. This allows for entire subsystems with expiring functionality to be replaced or upgraded, preserving the value of the remaining product. This type of "upward remanufacturing" [18] or "adaptability" [19] enables the remanufactured product to be incorporated into a new or "next generation" system.

Design for remanufacturing guidelines to assure future viability include [20, 21]:

- *Create a modular structure to make enable the product to keep pace with the changing needs of the end user through technical revitalization. As a result, the modular product may undergo several upgrades in components over its lifespan.*
- *Elements with short technology cycles should be designed to be easily upgradeable.*
- *Consider designs that allow software updates through connection without any disassembly.*
- *For aesthetic parts, avoid light colors, delicate surfaces, and enable easy replacement.*

Many product categories are also constrained by local, state, federal, and international regulations. These regulations can include energy efficiency requirements, material or chemical safety and use restrictions, takeback requirements, or emission limits to air, water, and land. As technology changes or new social issues arise, governments have the responsibility to react and either create new, or strengthen existing legislation to protect its citizens and/or natural resources. This regulatory creep can make products that are already on the market obsolete, limiting the remanufacturing of these products.

The premise of designing a product to be viable and compliant with future regulations is to be forward looking, and to consider the potential regulations that may be enforced at the time that the product is being remanufactured. One way to be forward looking is to evaluate the regulatory trend lines on the metrics of interest. Regulations such as automobile emissions or equipment energy efficiency continue to tighten as the product technology catches up with the current requirements. A forward looking product may try and design to meet the future regulation rather than just meeting the regulations currently being enforced.

Additionally, many material "red lists" identify materials that may not currently be regulated but are suspected of being hazardous. Tools such as the *chemsec SIN* (Substitute it Now!) list is a globally used database that identifies chemicals likely to be banned or restricted in a near future [22]. Designing to avoid future regulation may encompass the application of green chemistry practices to avoid materials on watch lists.

Finally, another option is to use new technologies that avoid the regulations altogether. For example, automobile manufacturers may avoid tightening emissions requirements by moving to electric or fuel cell propulsion technologies rather than internal combustion technology.

Another design strategy is to use modular design approaches that allow upgrade to modules that may be impacted by changes in regulation. In at least one regulated example, upgrades are not practical. In the case of heavy duty over the road truck engines, remanufactured engines are required to meet the original emissions specifications. Any changes to the engine design content would require recertification of emissions performance, possibly at new and stricter standards.

Design for remanufacturing guidelines to improve regulatory immunity include:

- *Know the local, national, and international environmental regulations governing the product before the design process begins. Project potential changes to those regulations that may occur over the product's life.*

- *Make an effort to design product and remanufacturing process to exceed the requirements of the current regulations. In the event regulation changes, this will avoid costly post-design phase modifications.*
- *Alternately, design the product with the consideration that certain modules or components may need to be upgraded to meet future regulation standards.*

6.4.2.3 *Design for Proactive Damage Prevention through Product Monitoring*

Designing to preserve value may include active monitoring of the product and its environment during use. Monitoring systems can be used to detect potential system or component failures and act on this information to prevent catastrophic failure, preserving the value of the system for future remanufacturing cycles. For an OEM, monitoring can support service decisions that maintain the value of the product during use, as well as to identify the appropriate time to remanufacture and to determine the value of a particular reman core [23, 24].

Design for remanufacturing guidelines to prevent damage through product monitoring include:

- *Incorporating warning systems for users to prevent damage due to over use or improper use. These systems can notify users regarding worn-out or malfunctioning parts, or scheduled service requirements. (Examples include: audible brake pad wear indicators, software update reminders, performance/service indicator lights)*
- *Integrate sensors and network connectivity to provide the manufacturer valuable data about the condition of the product to protect against product*

breakdowns that could result in loss of remanufacturing value.

6.4.3 Design to Recover Value

The final design for remanufacturing principle is to design the product so that the value is easily and cost effectively recoverable at the end of its use. The product and its components should be designed for ease of assessment of condition, so that decisions can quickly be made on the remanufacturing value, and to guide next steps in the process. The product should be designed for ease of disassembly and separation of modules and components, both to access and remove valuable components, and to enable further processing. Finally, the design should enable any required processing to bring the product back to the required like-new standards. Processing may include cleaning, material restoration, repair, or part replacement.

The design for remanufacturing literature has a strong focus on design to recover value. References that provide a good starting point for these design considerations include [9, 25–27]: The following subsections highlight design guidelines to improve recovery value.

6.4.3.1 Designing for Assessability

While component durability is an important goal of design, aging and damage mechanisms cannot be fully controlled. Further, different individual products may be exposed to different operating conditions and stressors. Because remanufactured products are expected to perform like new, it is imperative that components that cannot reliably perform over another use cycle can be identified. These components may be candidates for restoration, or may need to be recycled.

Designing for economic recovery starts with being able to accurately assess the value of the product when it is returned.

This allows remanufacturers to minimize handling and processing of low value cores, as well as having an understanding of the actual value of a core inventory. The second critical step is condition assessment of individual components for reuse or restoration. Some general design guidelines related to assessment are provided in Bras [20].

At the product level, having an understanding of the product usage history, or having means to identify severe exposure or abuse are important. This may include designing in of visual indicators that can identify severe exposure, for example to high temperature or to damaging environments. As an example, cell phones are designed with indicators of water and shock exposure. For overall product exposure, these indicators should be visible prior to product disassembly. A means of capturing life cycle histories of powered products, including overall operating hours, hours of operation at severe conditions, etc., can be enabled by embedded sensors or data storage devices such as an RFID readable tag.

For high value components, some of these same strategies can be used: designing in visual indicators of usage and overstress, and addition of sensors to capture specific usage history or condition related data. For examples, see [24] and [28].

Design for remanufacturing guidelines to enable product assessment include:

- *Develop indicators that will make it easier to evaluate the remaining life or stress exposure of cores. Some electronics, for example, use an indicator that changes color with moisture or severe temperature exposure.*
- *The highest form of inspection or tracking is a "smart product" or "smart part" that has tracked its usage and can relay this information at its end-of-life.*
- *Ensure that documentation and tracking mechanisms are available at product (and component) level, cradle to grave. This could include product*

documentation, markings or labels, bar code identification, or RFID technology.

- *All parts and modules that are remanufactured should be marked or tracked, so that the number of prior remanufacturing cycles can be identified for each part.*
- *Design product such that the high value components can be inspected with the fewest possible disassembly steps.*
- *Beyond usage tracking, products can also be designed to monitor sensors/signals that allow estimation of the condition and remaining useful life of high value modules or components. This enables improved disposition assessment at component and potentially product level.*
- *FMEA methods that consider failure modes and rates, and failure detection options, should be used when determining which components and failure modes/mechanisms are most critical to design for from an assessment standpoint.*

6.4.3.2 Designing for Separability/Disassembly (DfD)

Designing for economic recovery requires that the product can be easily disassembled in order to allow access and removal of valuable components and to enable further processing of the reman core. High value components should be removable without damage, and the time and effort to remove or separate high value components should be minimized.

In general, DfD involves selection of appropriate fastening and joining methods, and designing for access, and ease of component handling. This includes applying modular design principles, when possible, to group components with similar end-of-life disposition (e.g. reuse vs. recycle).

Methods for design for manufacture (DfM) and design for assembly (DfA) developed by Boothroyd and Dewhurst are widely applied in industry. Application of DfA principles can contribute to improved DfD, however design for assembly and disassembly principles can sometimes conflict. Boothroyd and Alting [29] recognize that separate criteria for DfD should be developed to support both serviceability and end-of-life material recovery.

Subramani and Dewhurst [30] evaluate DfD for service purposes and also have researched automated generation of product disassembly sequences [31]. Automated generation of disassembly sequences and process times has been a frequent research topic; and an overview of these studies was provided by Dong and Arndt [32]. The ability to automatically generate and analyze disassembly sequences provides a means to rapidly provide feedback on designs, but may not be very effective a providing guidelines or insight to drive design improvement.

Kroll and Hanft [33] describe a DfD scoring method similar to Bothroyd and Dewhurst's DfA approaches. Disassembly operations are given a base disassembly score depending on the type of operation, with difficulty scaling factors based on accessibility or ease of reach, the degree of positioning required, and the amount of force required for a task. These scores can be reasonably generated for a design prior to manufacturing involvement. A scoring system as described here is important during the design and development process for identification of potential pain points, and also for comparison with other products, and for tracking design improvement. When they are transparent to the designer, the scoring factors that are used in these processes can provide insight into potential design guidelines.

Design for recycling has been considered separately, and together with design for remanufacturing [34, 35]. Low cost separation of dissimilar materials is even more critical in recycling than remanufacturing, there is therefore significant overlap in design-for guidelines. In addition, effective recycling of materials that cannot be reused or restored is beneficial to remanufacturing operations. A major difference is that design for recycling can accommodate destructive disassembly, and separation processes, that are generally not applicable when considering design for remanufacturing.

Design for disassembly guidelines for remanufacturing, in addition to the above, include [20, 36, 37]:

- *Minimize the part count and part variety; standardize parts with common functions.*
- *Do not integrate parts/functions when they have fundamental differences in durability due to for example wear or fatigue.*
- *Avoid non-rigid parts.*
- *Design reusable parts to be stackable.*
- *Minimize the number of joints and connections, and make joints visible and accessible.*
- *Position the joints to minimize the need for realignment during dismantling. Minimize the number of handling operations that require heavy lifting.*
- *Protect joining elements from corrosion and wear.*
- *Avoid fastening methods that are not easily separable or cause damage in disassembly between modules or components that will be replaced, remanufactured or recycled.*
- *Minimize the need for specialized disassembly processes or tools.*
- *Minimize tight tolerances or narrow clearances on components that require disassembly or separation.*
- *Provide sufficient space between fasteners, away from obstructions and of easy access for the tools needed for disassembly.*
- *Standardize the fastener head technology used in the assembly and reduce the number of and variety of fasteners used.*
- *Isolate components or modules that may have material or handling hazards.*

- *Where possible, components in modules should have similar technology and physical (wear-out) obsolescence cycles.*
- *Minimize the number of connections between modules, as well as multi-modular connections.*
- *Minimize the number of disassembly operations required to separate components with different disposition (recycling versus remanufacturing)*
- *Use self-explanatory product structures, e.g. structures for which disassembly order, tools required, are obvious and/or clearly marked/identifiable/visible.*
- *Structure the product to make the highest value components the most accessible. Do not bury important components.*

6.4.3.3 Designing for Restorability

Finally, designing for economic recovery requires that the product components can be cost effectively restored to a like-new condition. In use, components can become contaminated with external materials such as dirt or oil, corrosion, or contaminants that are generated within a device such as wear particles or imaging toner. Reuse of contaminated components requires that they can be cleaned to like new condition. Cleaning is typically required prior to component condition assessment or restoration, and also is often required on external parts for aesthetic purposes. Aesthetic surfaces or finishes are often worn or damaged during use and must be restored.

Mechanisms of component aging can cause loss of performance or durability/reliability, for example: erosion or wear, mechanical fatigue, or change in material properties due to temperature exposure. In order to restore components that have experienced these types of aging, the design must consider cost effective processes for rebuilding or restoring functional surfaces (including re-machining or building up surface), repairing cracks (for example welding of high value castings), restoring fits between mating parts by adding or removing material (including sleeving or coating).

Very high level guidelines for cleaning and part reconditioning are provided in Hundal [35]. Bras [20] provides more detailed design guidelines, in particular with regards to cleaning. Yüksel [9] also provides some high level design guidelines related to cleaning as part of his QFD analysis for automobile engines. Ijomah [38] provides some additional design guidelines related to cleaning. A key issue during the design process is designing to accommodate available and cost effective cleaning processes. There is not significant coverage in the literature of cleaning processes as applied to remanufacturing, however Liu *et al.* [39] provide an overview of remanufacturing cleaning technologies and process considerations.

A summary of cleaning related design guidelines, drawn from the literature and from project experience at Rochester Institute of Technology, are provided below:

- *Part markings should withstand needed cleaning processes.*
- *Determine the aggressiveness of the remanufacturing cleaning process (time, temperature, chemistry, agitation) and choose materials that are stable, will not react, and will resist damage during the cleaning operation.*
- *Protect surfaces against corrosion and dirt adhesion to minimize required cleaning. Consider the types of potential contamination and the associated surface attraction forces; determine the best materials or surface treatments to minimize these contaminant adhesion. This may minimize the aggressiveness of the required cleaning process.*
- *Shield or protect high value modules or components from environmental contamination to minimize cleaning requirements.*
- *Minimize the need to mask parts to prevent damage or contamination during cleaning.*

- *Avoid product features that can be damaged by required cleaning or handling, such as delicate ribs or delicate materials or surface treatments.*
- *Minimize geometries that trap contaminants over the service life or limit access to cleaning processes, for example small fluid passages or blind holes, reentrant corners and undercuts.*
- *Minimize cavities that may collect cleaning residues (contaminants or cleaning media) during cleaning processes.*
- *Do not use surface treatments or assembly processes that cannot be readily reversed and potentially limit access for component replacement/repair.*
- *Consider how parts will be fixtured for effective cleaning.*

A summary of restoration related design guidelines based on Bras [20] and project experience at Rochester Institute of Technology, are provided below:

- *Design product so that setting adjustments are easily accessible and clearly marked.*
- *Special consideration must be taken for surfaces with high tolerance requirements or that are subject to wear. Remachining may be needed, leading to the requirement to have additional material thickness in the original product or means of building up surfaces.*
- *Design mechanical components for infinite fatigue life, or allow for ease of separation of life-limited components.*
- *Likewise, components with tight assembly fits require a mechanism for reestablishing production fit tolerances (sleeving or coating bores, oversizing pistons or rods, etc).*
- *Design wear interfaces such that the wear is concentrated on easily repairable or low cost replaceable components.*

- *Similarly, corrosion can be difficult to restore in components, therefore select materials/pairs that minimize corrosion or concentrate corrosion on low cost replaceable components.*
- *Special surface treatments must be durable, or readily repairable or replaceable.*
- *Surfaces with aesthetic function should be durable or easily restorable (including cleaning without damage). Consider cleaning requirements when selecting finishes such as paint or coatings or platings.*

6.5 Conclusion

Product design is a very important determinant in the feasibility (technical and economic) of product remanufacturing. Three major principles to designing for remanufacturing have been described: (1) design to create value, (2) design to protect and preserve the value, and (3) design to easily and cost effectively recover the value. Design to create value is inherent in all product design although the guidelines provided, particularly with respect to integrating value, can have a greater benefit for companies planning to remanufacture. Design to recover value guidelines have received more attention in previous analyses of design for remanufacturing. While remanufacturers have learned to be innovative in order to overcome poor design practices, good design is an enabler of more cost effective remanufacturing. Finally, failure to design to protect and preserve value is a significant barrier to the market potential of remanufactured products and therefore to growth of the remanufacturing industry.

Companies such as Caterpillar and Xerox have made remanufacturing a cornerstone of their efforts to provide the highest value products to their customers. Companies that are planning to remanufacture their own products have a strong incentive to consider the recommendations presented here within their design processes. However design trade-offs with respect to cost and performance are already challenging for product developers. In addition to consideration of these directional guidelines, processes to facilitate design trade-offs that include remanufacturing criteria are also needed.

In some industry sectors, such as automotive, more of the remanufacturing is done by third party companies than Original Equipment Manufacturers, therefore the incentives to include remanufacturing criteria in the original design process are not as strong. However at least for durable products, there is a need to provide service parts that OEMs must take into consideration. In many cases there are OEM relationships with remanufacturers to support service demands in order to provide reasonably priced parts for customers. Increased consideration of remanufacturing during the original design process will facilitate cost effective remanufactured service parts. Many of the product design criteria for remanufacturing also are directionally correct for improved maintainability and reparability which also should play a role in product design.

6.6 Acknowledgements

Many of the insights shared in this chapter are, in part, the fruit of extensive research conducted by the Golisano Institute for Sustainability (GIS) at the Rochester Institute of Technology (RIT), under the United States National Institutes for Standards and Technology (NIST) Advanced Manufacturing Technology Consortia (AMTech) grant program. Associate Provost and Director of the Golisano Institute for Sustainability Dr. Nabil Nasr and Senior GIS Research Engineer Brian Hilton led significant efforts in industry outreach, collaborative workshopping, and technology roadmapping that provided valuable industry perspectives on critical factors in the growth of the U.S. remanufacturing industry. These efforts were supported by Victoria Brun and Dr. Frederick Hanson of Energetics, Inc., the

Remanufacturing Industries Council (RIC), and innumerable subject-matter experts and industry representatives who volunteered their time to support the advancement of the industry. Special thanks is extended to all those who contributed.

References

1. United Nations Department of Economic and Social Affairs, Population Division. World Population Prospects, 2015. Revision. https://esa.un.org/unpd/wpp/.
2. U.S. Department of Commerce, U.S. and World Population Clock. Accessed March 21, 2017. https://www.census.gov/popclock/.
3. UNU-IHDP and UNEP, *Inclusive Wealth Report 2012: Measuring Progress toward Sustainability*, Cambridge University Press, New York, NY, 2012.
4. UNEP, 2010. Assessing the Environmental Impacts of Consumption and Production: Priority Products and Materials. A Report of the Working Group on the Environmental Impacts of Products and Materials to the International Panel for Sustainable Resource Management. Hertwich, E., van der Voet, E., Suh, S., Tukker, A., Huijbregts M., Kazmierczyk, P., Lenzen, M., McNeely, J., Moriguchi, Y.
5. Nasr, N., Hilton, B., German, R., A framework for sustainable production and a strategic approach to a key enabler: Remanufacturing, in: *Advances in Sustainable Manufacturing*, G. Seliger, M. Khraisheh, I. Jawahir (Eds.), pp. 191–196, Springer, Berlin Heidelberg, 2011.
6. United States International trade Commission, *Remanufactured Goods: An Overview of the U.S. and Global Industries, Markets and Trade*, USITC Publication 4356, Washington D.C, 2012.
7. Kushma, D., Jones: Built the nation's largest parts remanufacturing business, Automotive News, November 2009, "Jones' operation grew to occupy nine buildings on more than six blocks in downtown Oklahoma City. It provided rebuilt engines and components to 2,500 dealerships." https://www.autonews.com/article/20091109/RETAIL/311099741/jones-built-the-nation-s-largest-parts-remanufacturing-business

8. Akao, Y. (Ed.), *Quality Function Deployment: Integrating Customer Requirements into Product Design*, (Translated by Glenn H. Mazur) Productivity Press, Cambridge, MA, 1990.
9. Yüksel, H., Design of automobile engines for remanufacture with quality function deployment. *Int. J. Sustain. Eng.*, 3, 3, 170–180, 2010.
10. Anderson, D., *Design for manufacturability & concurrent engineering: How to design for low cost, design in high quality, design for lean manufacture, and design quickly for fast production*, CIM press, Cambria, CA, 2004.
11. Shimbun, N.K., *Poka-yoke: Improving product quality by preventing defects*, Productivity Press, New York, NY, 1989.
12. Taguchi, G., Chowdhury, S., Taguchi, S., *Robust Engineering*, McGraw-Hill Professional, New York, NY, 2000.
13. Salhieh, M.S. and Kamrani, A.K., Macro level product development using design for modularity. *Robot. Comput. Integr. Manuf.*, 15, 4, 319–329, 1999.
14. Sanderson, S.W. and Uzumeri, M., *Managing Product Families*, Irwin Professional Pub, New York, NY, 1997.
15. King, A.M. and Burgess, S.C., The development of a remanufacturing platform design: A strategic response to the Directive on Waste Electrical and Electronic Equipment. *Proc. Inst. Mech. Eng. B J. Eng. Manuf.*, 219, 8, 623–631, 2005.
16. Bras, B. and Hammond, R., Towards design for remanufacturing—Metrics for assessing remanufacturability. *Proceedings of the 1st International Workshop on Reuse. Eindhoven, Netherlands*, 1996.
17. Scaramella, J., Daly, J., Marden, M., Perry, R., The Cost of Retaining Aging IT Infrastructure, IDC White Paper #246755, International Data Corporation, 2014.
18. Nasr, N. and Thurston, M., Remanufacturing: A key enabler to sustainable product systems. *Proceedings of 13th CIRP International Conference on Lifecycle Engineering*, 2006.
19. Li, Y., Xue, D., Gu, P., Design for product adaptability. *Concurr. Eng.*, 16, 3, 221–232, 2008.
20. Bras, B., Design for remanufacturing processes, in: *Environmentally Conscious Mechanical Design*, M. Kutz (Ed.), pp. 283–318, John Wiley and Sons, Hoboken, NJ, 2007, ISBN 978-0-471-72636-4.

21. Bras, B., Product design issues, in: *Closed-Loop Supply Chains: New Developments to Improve the Sustainability of Business Practices*, M. Ferguson, Souza, G. (Eds.), Auerbach Publications, New York, NY, 2010.
22. chemsec SIN List, *The International Chemical Secretariat*, 2017, http://sinlist.chemsec.org/.
23. Sundin, E. and Lindahl, M., Rethinking product design for remanufacturing to facilitate integrated product service offerings. *2008 IEEE International Symposium on Electronics and the Environment*. San Francisco, CA, pp. 1–6, 2008.
24. Klausner, M., Grimm, W.M., Hendrickson, C., Horvath, A., Sensor-based data recording of use conditions for product takeback. *Electronics and the Environment. 1998. Proceedings of the 1998 IEEE International Symposium on*, pp. 138–143, 1998.
25. Berko-Boateng, V., Azar, J., De Jong, E., Yander, G.A., Asset cycle management-a total approach to product design for the environment. *Electronics and the Environment, 1993. Proceedings of the 1993 IEEE International Symposium on*, pp. 19–31, 1993.
26. Sundin, E. and Bras, B., Making functional sales environmentally and economically beneficial through product remanufacturing. *J. Clean. Prod.*, 13, 9, 913–925, 2005.
27. Hatcher, G.D., Ijomah, W.L., Windmill, J.F.C., Design for remanufacture: A literature review and future research needs. *J. Clean. Prod.*, 19, 17, 2004–2014, 2011.
28. Kara, S., Mazhar, M., Kaebernick, H., Ahmed, A., Determining the reuse potential of components based on life cycle data. *CIRP Ann. - Manuf. Technol.*, 54, 1, 1–4, 2005.
29. Boothroyd, G. and Alting, L., Design for assembly and disassembly. *CIRP Ann. - Manuf. Technol.*, 41, 625–636, 1992.
30. Subramani, A. and Dewhurst, P., Efficient design for service considerations. *Manuf. Rev.*, 6, 1, 40–47, 1993.
31. Subramani, A.K. and Dewhurst, P., Automatic generation of product disassembly sequences. *CIRP Ann. - Manuf. Technol.*, 40.1, 115–118, 1991.
32. Dong, J. and Arndt, G., A review of current research on disassembly sequence generation and computer aided design for disassembly. *Proc. Inst. Mech. Eng. B J. Eng. Manuf.*, 217, 3, 299–312, 2003.

33. Kroll, E. and Hanft, T.A., Quantitative evaluation of product disassembly for recycling. *Res. Eng. Des.*, 10, 1, 1–14, 1998.
34. Kriwet, A., Zussman, E., Seliger, G., Systematic integration of design-for-recycling into product design. *Int. J. Prod. Econ.*, 38, 1, 15–22, 1995.
35. Hundal, M., Design for recycling and remanufacturing. *Proceedings of International design conference - design 2000.* Dubrovnik, Croatia, 2000.
36. Bogue, R., Design for Disassembly: A Critical Twenty-First Century Discipline, *Assembly Autom.*, 27.4, 285, 2007.
37. Warnecke, H.J., Schwelzer, M., Kahmeyer, M., Flexible disassembly with industrial robots. *Proceedings of the International Symposium on Industrial Robots*, International Federation of Robotics, & Robotic Industries, vol. 23, pp. 547–547, 1992.
38. Ijomah, W.L., Addressing decision making for remanufacturing operations and design-for-remanufacture. *Int. J. Sustain. Eng.*, 2, 2, 91–102, 2009.
39. Liu, W., Zhang, B., Li, M.Z., Li, Y., Zhang, H.C., Study on remanufacturing cleaning technology in mechanical equipment remanufacturing process, in: *Re-engineering Manufacturing for Sustainability*, pp. 643–648, 2013.

General References

IRP, Re-defining Value – The Manufacturing Revolution. Remanufacturing, Refurbishment, Repair and Direct Reuse in the Circular Economy. Nabil Nasr, Jennifer Russell, Stefan Bringezu, Stefanie Hellweg, Brian Hilton, Cory Kreiss, and Nadia von Gries. A Report of the International Resource Panel. United Nations Environment Programme, Nairobi, Kenya, 2018.

Ijomah, W.L., McMahon, C.A., Hammond, G.P., Newman, S.T. Development of robust design-for-remanufacturing guidelines to further the aims of sustainable development. *Int. J. Prod Res.*, 45, 18–19, 4513–4536, 2007.

Nasr, N., Becker, M., Haselkorn, M., Jessop, S., Thorn, B., LaRochelle, M. Closing the Loop: Design Tools for Sustainable Products, Report to U.S. EPA, Washington, D.C., 2002.

Nasr, N., Berg, J., Cleveland, J., Hilton, B., Roth, B., Scheldorn, J., Slocum, A., Sustainable Design Tools for Proton Membrane Exchange Fuel Cells, Report to U.S. EPA, 2004.

Nasr, N. and Hilton, B., Design for Remanufacturing. *In Proceedings 15th CIRP International Conference on Life Cycle Engineering*, Sydney, 2008.

National Center for Remanufacturing and Resource Recovery 2005, The U.S. Remanufacturing Industry, Terms & Definition.

7

Global Challenges and Market Transformation in Support of Remanufacturing

Shanshan Yang

Advanced Remanufacturing and Technology Centre, CleanTech Two, Singapore

Abstract

In an era where demand for finite resources will only continue to rise, the way people produce and consume goods must change; in short, the global community must transition from linear systems of extraction, production, consumption, and disposal to a more circular economy. As an effective closed-loop measure in circular economy, remanufacturing presents immense potential for economic, environmental, and social benefits. However, global uptake of remanufacturing is still hampered by a number of challenges which will be analyzed in this chapter. To address these barriers is simply beyond any single entity's scope of capability; rather, large scale adoption of remanufacturing in pursuit of a circular economy requires collaboration amongst multiple players across business, research, government, and investor communities. In this chapter, key areas that will catalyse the industry and drive the scale-up of remanufacturing are identified and presented. Successful business cases are also included to showcase accelerated adoption of remanufacturing model.

Email: yangs@artc.a-star.edu.sg

Nabil Nasr (ed.) Remanufacturing in the Circular Economy (169–210)

Keywords: Challenge of remanufacturing, remanufacturing market, circular economy, product service system, reverse supply chain, design for remanufacturing, industry 4.0 and remanufacturing

7.1 Introduction

Traditional linear industrial models of "take-make-consume-dispose" ideology are increasingly challenged by an unprecedented rise in demand for finite resources associated with both population growth and the resultant desire for widespread socioeconomic development. In response, the circular economy is emerging as one of the key strategies by which global economies may decouple this desired (and required) economic growth from unsustainable consumption of constrained resources. In the circular economy, an altogether different philosophy as championed as the precipitator of success—parts or products (i.e. technical materials) are deliberately designed to repeatedly circulate within continuously regenerative loops of material flow, allowing industrial and economic growth to become detached from their longstanding reliance on unfettered and perpetual access to new raw materials. Beyond design, key actions are required at the end of product life to enable more circular systems. Specifically, as elaborated in Figure 7.1, reusing, remanufacturing, or recycling both whole products and their constituent components minimizes consumption of raw materials. Among these End-of-Life (EOL) strategies, remanufacturing shows notable

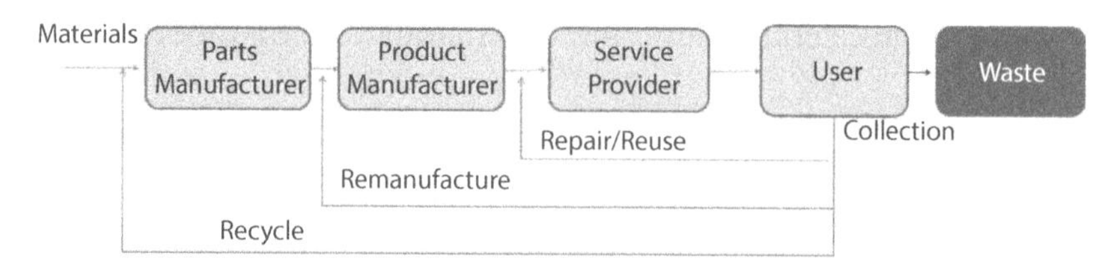

Figure 7.1 Technical materials in a circular economy system.

advantages in closing the loop of material flow, preserving added-value in products, and ensuring the quality of a product's next life cycle.

Remanufacturing is the process of bringing products back to sound working status through disassembly, cleaning, sorting, inspection, reconditioning and reassembly. It has demonstrated promising potential for creating social, economic, and environmental benefits well beyond what can be realized through simple repair or base material-level recycling. These benefits include decreasing the costs expended and resources exploited during production process, protecting Intellectual Property and brand image for Original Equipment Manufacturers (OEMs), creating new business opportunities in the after-sales service market, and creating new job opportunities for skilled laborers [1].

Despite the evident potential for remanufacturing to enable the circular economy, global uptake of remanufacturing remains slow, still hampered by a number of challenges. Trade laws in many countries, for example, still fail to differentiate remanufactured or remanufacturable products from used products at EOL, causing them to restrict the import of cores or products in effort to combat the international trade of EOL products that are effectively waste. Likewise, product design, core supply, and market demand also slow the momentum of remanufacturing.

As the challenges faced by remanufacturers vary among countries and industry sectors, the first half of this chapter provides an overview of the regional remanufacturing landscape and individual product sectors. Understanding this context will thus support more comprehensive global views of challenges faced by remanufacturers worldwide, discussed in the second half of this chapter. This understanding then informs a discussion of the key enablers required to create new opportunities for remanufacturing, supported by case studies of successful business applications that showcase accelerated adoption of remanufacturing and its potential benefits.

7.2 Global Remanufacturing Landscapes

7.2.1 The United States

The United States is undisputedly the leader in terms of producing, consuming and exporting remanufactured goods, where estimates suggest that remanufactured goods account for nearly two percent of all product sales [2]. In 2011, the value of U.S. remanufacturing production reached $43.0 billion, growing 15 percent since 2009, with international exports accounting for $11.7 billion of that amount. Canada, the European Union (EU), and Mexico were the leading destinations for U.S. exports. In addition to the major OEM remanufacturers, small- and medium-sized enterprises (SMEs) generated 25 percent ($11.1 billion) of remanufacturing productivity, and are responsible for 36 percent of the more than 180,000 full-time jobs in U.S. remanufacturing, illustrating the importance of supporting remanufacturing in a variety of business contexts. This vitality is based on the prevalence of remanufacturing business models in key industrial sectors, including aerospace, consumer products, electrical apparatus, heavy-duty and off-road (HDOR) equipment, information technology (IT) products, locomotives, machinery, medical devices, motor vehicle parts, office furniture, restaurant equipment and retreaded tires [3]. Among these, aerospace, HDOR, and motor vehicle parts sectors account for 63 percent of total remanufacturing production (Table 7.1) highlighting both their viability within and criticality to the success of circular industrial systems.

7.2.2 Europe

Europe is also a major player in the global remanufacturing industry, with a strong focus on aerospace, automotive and HDOR sectors. Estimates suggest that European remanufacturing productivity generated approximately $42 billion, accounting for nearly 1.9% of product sales and employing 190,000

Table 7.1 Estimation of remanufactured goods production, employment, and exports in the US—2011 [4].

Sector	Examples of products	Production (thousands $)	Employment (full-time workers)	Exports (thousand $)
Aerospace	Engine, aircraft's airframe	13,045,513	35,201	2,589,543
HDOR equipment	trucks, bulldozers, excavators etc.	7,770,586	20,870	2,451,967
Motor vehicle parts	passenger cars and light trucks etc.	6,211,838	30,653	581,520
Machinery	Marching tools	5,795,105	26,843	1,348,734
IT Products	personal computers (PCs), servers, Motherboards, and hard disk drives.	2,681,603	15,442	260,032
Medical devices	medical imaging equipment	1,463,313	4,117	488,008

(*Continued*)

Table 7.1 Estimation of remanufactured goods production, employment, and exports in the US—2011 [4]. (*Continued*)

Sector	Examples of products	Production (thousands $)	Employment (full-time workers)	Exports (thousand $)
Retreaded tires	types of trucks and buses, heavy construction and agricultural equipment, aircraft, and passenger vehicles	1,399,088	4,880	18,545
Consumer products	Washing machine, air conditioner	659,175	7,613	21,151
Others	Furniture, Restaurant equipment	3,973,923	22,999	224,627

people in 2011. Germany, the United Kingdom & Ireland (UK), France, and Italy are four key nations that hold critical roles in this region as top producers and leaders in technology. The collective value of their remanufacturing industries was estimated to account for 70% of the total European remanufacturing market. Most EU members view remanufacturing as both a sustainable business model and a means to protect the environment. Accordingly, several mandatory legal directives have been issued to foster the growth of remanufacturing in Europe. Examples include management directives for End-of-Life Vehicles (ELV) and Waste Electrical and Electronic Equipment (WEEE), which set targets for recycling and reuse in automotive and electrical/electronic sectors respectively, and thus encourage remanufacturing activities through economic incentive. With proper governmental support and promotion, the European remanufacturing industry is expected to grow to approximately $120 billion per year and add nearly 65,000 jobs by 2030.

7.2.3 China

The total value of the remanufacturing industry in China was estimated to reach $20 billion in 2015 [5]. The growth of the Chinese remanufacturing industry is primarily fostered by government regulation that aims to reduce environmental pollution and promote sustainability. In this pursuit, the State Council of China officially supports for remanufacturing industry as a key element of the Chinese economy. Since their original endorsement in 2005, more than 20 laws have been passed to expand the applicability of remanufacturing. Pilot programs in this pursuit aim to explore the potential for integrating remanufacturing as an aspect of new good production. For example, some motor vehicle parts manufactures can remanufacture engines, transmissions, generators, starters, drive shafts, compressors, oil pumps, water pumps, and other components that may be incorporated into effectively new cars [6]. As remanufacturing activates have only begun over the last decade in China, there are still many barriers that limit its growth. Definition and standards for remanufacturing, for example, are still lacking. As a result, remanufactured or remanufacturable goods are often classified as "old" or "scrap" products, causing international trade in goods and cores to be restricted or prohibited.

7.2.4 Other Countries

Remanufacturing in other countries is relatively limited scope and value. Relevant progress is identified in in Table 7.2 [4].

7.3 Overview of Remanufacturing Sectors

Over years of development, remanufacturing has become valuable practice across a number of industries. Typically, remanufacturing has a higher chance for long-term economic viability in industries where products are durable, have high value, and use technology that is stable over more than one life cycle. In addition, the availability and accessibility of restoration technologies, as well as the ability to employ product-service-system business models, are also key strengthening attributes for remanufacturing business models. For these reasons, remanufacturing has become relatively well-adopted in sectors such as aerospace, automotive, and HDOR, where these system characteristics had long presented quandaries in the new-product-reliant business model. Today, remanufacturing activities in these sectors are notably more established and account for over 60% of the remanufacturing industry's production value. The rise of remanufacturing due more explicitly to emerging environmental awareness and subsequent regulatory scrutiny has been slower—but is gaining momentum—in industries such as IT products, machinery, retreaded tires, where the economic opportunity required greater exploration and careful definition in light of successful incumbent business models.

Table 7.2 Remanufacturing activities and related measures in selected countries.

Country	**Remanufacturing sectors**	**Measures affecting remanufacturers**
Singapore	Aerospace, Machinery, Marine, HDOR	• Free trade agreement in Association of Southeast Asian Nations (ASEAN) countries and Australia. • Advanced Remanufacturing and Technology Center, an R&D center for remanufacturing technology development
Brazil	Aerospace, Motor Vehicle Parts, HDOR, IT products	• Government does not distinguish used from remanufactured. • Restrictions prohibit import of cores & remanufactured goods
India	IT products and HDOR	• Restriction on foreign trade of remanufactured goods & cores
Canada	Motor Vehicle, Aerospace, Medical	• Government does not distinguish new from remanufactured

(*Continued*)

Table 7.2 Remanufacturing activities and related measures in selected countries. (*Continued*)

Country	**Remanufacturing sectors**	**Measures affecting remanufacturers**
Malaysia	Motor Vehicle, IT, Aerospace, Printers	• Used parts prohibited from import; lmits core availability
Mexico	Consumer Electronics, Printers & Cartridges, Motor Vehicle Parts	• No specific laws available to regulate remanufacturing activity. • "Three Rs" (reduce, reuse, and recycle) initiative to promote sustainability and reduce industrial waste
Japan	Imaging, Aerospace, HDOR	• No restriction on export and import of remanufactured goods (aerospace, medical devices, and tire)

7.3.1 Aerospace

Aerospace remanufacturing is driven by the demand for continuous maintenance and overhaul service often widely regulated and required by aircraft in commercial, private, and military fleets. Systems and subsystems frequently remanufactured include airframes, engines, avionics, hydraulics, and interior furnishings. Smaller system components with measurable wear characteristics, called "rotables," are often subject to defined service periods after which they must be remanufacture. These products include wheels, brakes, auxiliary power units, fuel systems, flight controls, thrust reversers, landing gear assemblies, and electrical systems.

Remanufacturing in this context covers a wide range of activities, including repairing used components, replacing worn components with virgin (new) parts, and upgrading product performance maintain parity with current technologies. In fact, the terms "overhaul," "rebuild" and "repair" are more frequently used to denote "remanufacturing" in the aerospace industry. In many, if not most cases, the parameters of remanufacturing efforts—from the duration of use until remanufacturing is required to the functional performance requirements after processing—are defined by tightly-controlled safety and quality regulations. Traditionally, OEMs have little involvement in the maintenance, repair and overhaul market, and independent remanufacturers are therefore often left struggling to efficiently obtain design and performance information to ensure remanufactured products adequately meet regulated specifications. As the maintenance and repair market grows, however, OEMs are increasingly offering service-based purchasing agreements to gain advantages in aftermarket. Airbus, for example, offers Flight Hour Services in which an airline can select the level of support it needs to complement its own maintenance, reducing reliance on independent parts remanufacturers.

Aerospace remanufacturing is also challenged by the availability of skilled and certified technicians. The pool of qualified workers is still relatively limited. Meanwhile, advances in technology that require advanced skills—such as new composite materials or fuel cell power systems—hold great potential to generate drastic impact on the aerospace sector (and thus, inherently, its remanufacturing industry) but are struggling to gain traction due to this deficit.

7.3.2 Automotive Parts

Automotive parts remanufacturing is reported as the world's largest remanufacturing sector [7]. Frequently remanufactured products include engines, transmissions, starter motors, alternators, steering racks, and clutches. With the gradual increase in vehicle longevity, the average life of vehicles on the road is now over seven years. Thus, because cars are on the road longer, the demand for spare parts is increasing, creating opportunity for growth in the automotive remanufacturing sector. In Europe, the End-of-Life Vehicle (ELV) Directive has further buoyed the remanufacturing industry in this sector by dictating a percentage of end-of-life (EOL) vehicles that must (or, rather, should) be reused or recycled, providing economic incentive for both OEMs and remanufacturers to circularize business models and reduce the amount of waste generated when vehicles reach their end-of-use stage.

The challenges faced by automotive parts remanufacturers come from high labor cost (due to product complexity), scarcity of quality and adaptable cores, and increasing design complexity in which electrification and electro-mechanical integration are making vehicles more difficult to cost-effectively assess, diagnose, disassemble, and repair. Compared to virgin (new) part manufacturing, the disassembly, separation, cleaning, and repair operations required in remanufacturing involve three times as much labor [4]. This creates high labor costs that, combined with competition from low-cost virgin alternatives, have made remanufacturing an economically unattractive option in some cases.

The availability of quality cores also hampers the growth of remanufacturing. Estimates suggest that the cost of core can represent anywhere from 19 to 59 percent of a remanufactured product, implying that the benefit of remanufacturing operations is largely affected by the quality of the core collected. Because the use environments, intensities, and conditions of vehicles are so widely variable, this creates a degree of unpredictability in core quality, subsequently requiring remanufacturers to constantly adjust and limiting the potential for standardized, streamlined processes that achieve greater cost efficiency. In addition, driven by ever-evolving emissions regulations, engine technology is changing every two to three years. As a result, engine designs are becoming increasingly complex and integrated with electronics, adding to the complexity of remanufacturing operations and requiring more advanced skillsets.

7.3.3 Heavy-Duty and Off-Road (HDOR)

Remanufacturing is particularly prevalent in the HDOR sector, as large, complex, capital intensive, and durable products are well-suited to repeated service and life cycle extension rather than outright replacement. Remanufacturing activities in this sector focus on vehicles such as trucks, bulldozers, excavators, backhoes, asphalt pavers and rollers, farm tractors, combines, rock cutters, tunnelling machinery, and oil and gas drilling machinery. The high-value industries in which such equipment operates often means that downtime can be detrimental; as a result, repair and remanufacturing in the field are emerging as cost-effective strategies. Similarly, due to the high capital costs of new HDOR equipment, remanufactured products may cost 70% less than new products. Accordingly, recognizing the potential appeal of remanufacturing to customers (and thus its economic opportunity), an estimated one third of OEMs in the HDOR sector produce both new and remanufactured equipment.

Similar to automotive remanufacturing, the HDOR sector also faces challenges in high labor cost, competition from cheap raw materials for new product manufacturing, and increasing electrification of vehicle systems. In addition, because HDOR equipment and its components are expectedly diverse in design (as a natural result of a multi-competitor market), the relative non-standardization and/or interchangeability of parts and components often necessitates the maintenance of part inventories to cover all potential replacement scenarios. In some ways, this creates a challenge for remanufacturers, who may have difficulty both finding replacement parts for legacy equipment and meeting demand for contemporary replacement parts with existing inventories. In this sense, much technical alignment and organization is needed.

7.3.4 Information Technology (IT)

Information technology (IT) is perhaps one of the most challenging sectors for remanufacturing. Commonly remanufactured products include personal computers (PCs), servers, motherboards, circuit packs, modems, and office technologies such as printers and toner cartridges. In this field, "refurbished" is a more common word than "remanufacturing" and involves disassembly, cleaning, inspection, and replacement of faulty components with functioning parts. Units are then tested to ensure electrical safety and functional performance. Comparing with independent remanufacturers, OEMs are reported to account for a larger share of this activity due to the highly complex and proprietary nature of most IT products. Despite this, remanufacturing only accounts for a small portion of OEM revenues. Third party remanufacturers may still acquire, remanufacture, and sell products independently (and thus pose some small degree of direct competition with OEMs), but few, if any, independent remanufacturers are certified by the OEM, often impacting customer perceptions of quality and performance.

Compared to other sectors, electronics and IT products face a unique set of challenges. The rapid advancement of electronic and digital technologies renders many products effectively obsolete in as little as two to three years. In many cases, such older products cannot support upgrades that achieve parity with contemporary market offerings, and thus do not fulfil the potential to compete with current virgin (new) products. In addition, because of this rapid turnover, consumers typically seek to balance the most advanced functionality with the most economical solutions, creating difficulty forecasting supply and demand for remanufactured products. Beyond this, many consumers either do not know how or are reluctant to send their IT products for remanufacturing, due either to a lack of knowledge about options or concerns regarding data security. As a result, experts find that the supply of quality cores is insufficient to support widespread and viable remanufacturing in this industry [8]. In response (at least in Europe) the Waste Electrical and Electronic Equipment (WEEE) Directive has extended producer responsibility for managing the product wastes at EOL and set targets for recovery and recycling, which has further encouraged remanufacturing. Troublingly, in cases where technology does not advance as rapidly, OEMs worried about remanufacturing cannibalization of new product sales often deliberately hamper remanufacturing activates by discrediting the quality of third-party remanufactured products, promoting the use of new products, or even designing the product to more complex specifically to complicate or inhibit remanufacturing.

7.3.5 Other Sectors

Remanufacturing also has its presence in other sectors. Their respective challenges are explained in Table 7.3.

Table 7.3 Characteristics of remanufacturing in selected industry sectors [4, 8].

Sector	Products remanufactured	Challenges
Medical Devices	Single-use devices, imaging equipment, portable ultrasound	• Core quality • Lack of suitable dealers and distribution networks • Legislative restrictions
Furniture	Office chair, lateral file, desks, office cubes	• Customer perceptions/ OEM fears • Core collection/reverse logistics • Legislative restrictions
Machinery	Machine tools, pumps, compressors, engines, turbines, food processing	• Lack of remanufacturing option awareness • High core/transportation costs • Availability of skilled workers
Marine	Engine and transmission	• Non-standardized equipment and components • Uncertain timing and quality of core return • Lack of knowledge, skills, and yard-capacity
Retreaded Tires	Tires for trucks, buses, construction and agricultural, aircraft, passenger vehicles	• Core availability • Poor customer perceptions • Competition from cheap new tires

7.4 Global Challenges

Because remanufacturing industries have grown at a global scale, barriers to remanufacturing are felt in different ways by players around the globe depending on myriad geopolitical, economic, and environmental factors. In addition, the diversity of models for handling business, intellectual property rights, and data accessibility often means these challenges are also experienced differently between business types, with OEMs and third-party remanufacturers facing different sets of challenges in many contexts. Across this spectrum, however, there remain some common challenges shared almost ubiquitously.

7.4.1 Standards & Legislation

The absence of a commonly accepted legal definition for remanufacturing and associated certifications for remanufactured goods is cited as one of the most prevalent barriers for remanufacturing [3]. In many cases, remanufactured goods are frequently classified, in terms of international trade, as "used." The impacts of this poor distinction are more than simply semantic; many nations restrict the import and export of "used" products in effort to minimize the trade of EOL products that are effectively waste material, and thereby avoid the economic and environmental costs of managing that waste. In some cases, such restrictions reflect efforts to protect domestic markets for secondary life cycles. In precluding the trade of remanufactured or remanufacturable goods, however, such policies constrain the growth of remanufacturing at larger scales. This barrier is especially apparent in China and India, where population growth and, therefore, market potential are both immense and growing.

At a smaller scale, the lack of definition and standards for remanufacturing has also led to inconsistency across markets in the quality and performance of remanufactured products as compared to virgin (new) counterparts. Some companies sell products as "remanufactured" in effort to capitalize upon the term's connotation of performance even though the products are in fact only reused or repaired. Customers purchasing such falsely-branded products and quickly finding malfunction may thereafter perceive remanufactured products in general as "low quality." This inconsistency, then, creates variable—but often negative—consumer trust in remanufactured products as a whole, putting companies engaged in true, high-quality (i.e. new-equivalent) remanufacturing at significant disadvantage. In this regard, a widely accepted definition of remanufacturing, a standard for inspection, restoration and testing, and a model for new-equivalent performance certification is needed [3].

Over the decades, there have been a number of laws and regulations to support EOL product disposition, such as the End-of-Life Vehicles Directive, the Waste Electrical and Electronic Equipment (WEEE) Directive, and Restriction of the Use of Certain Hazardous Substances (RoHS). However, the impact of these regulations varies. While they set realistic targets for reuse, recycling, and recovery and extend producer responsibility more completely over the product life cycle, they can also limit remanufacturing in some ways. For example, RoHS regulates the amount of lead (Pb) that may be used in electronics and electronic components in the EU market. While the environmental and human health intentions are clear, this may stop remanufacturers from reusing components that contain lead, forcing them to dispose of such components (which creates environmental exposure anyway) and replace them with lead-free parts (driving up remanufacturing cost and creating even further economic barriers). As another example, specific performance standards for remanufactured devices in the medical industry are necessary, but impose additional costs that discourage remanufacturing in favor of repair, which does not require recertification [8]. This not only stifles the growth of remanufacturing, but also means that healthcare providers often settle for equipment that functions only repair-to-repair, creating frequent downtimes that affect both human health and economic efficiency.

7.4.2 Design

Design plays a critical role in determining the potential for remanufacturing a product. The lack of specific design considerations for remanufacturing creates many of the challenges encountered during remanufacturing processes today. For example, use of permanent joints will add complexity to disassembly process, sometimes requiring destructive disassembly methods that can cause unnecessary damage to neighbouring parts, ultimately creating losses in the product's recoverable value. Another example is the non-standardization of part and component designs across manufacturers, and even across different products from a single manufacturer. While this is often a natural result of multi-player competition within a market, it often creates diversity that requires extra effort during parts assessment and sorting processes.

Currently, few companies (except for major players such as Xerox, Caterpillar and Kodak) have desire to actively enhance product remanufacturability through design. In some cases, companies simply lack awareness of the potential benefits of designing for remanufacturing. In others, companies often fear that designing for remanufacturing may benefit independent remanufacturers, who ultimately become strong competitors in the aftermarket, more than the OEM itself. In fact, some OEMs even deliberately design products to inhibit or complicate remanufacturing explicitly to combat the loss of business to third-party remanufacturing. Some printer ink cartridge OEMs, for example, have implanted microchips in their products that render them inoperable if 'tampered' with. Though the intention of this was apparently to prevent counterfeiting, such features also—and probably not accidentally—effectively destroy all functional value if disassembled for remanufacturing.

Compared to other strategies such as design for manufacturing or performance, design for remanufacturing is usually not an issue for a typical designer because the corporate philosophy driving the design either ignores or specifically excludes it. Due to this lack of consideration and an over-emphasis on cost-effectiveness, remanufacturers often struggle to process returned products due to their poor quality, or have difficulty finding economic justification because the potential recoverable value does not sufficiently outweigh the extensive work required to remanufacture a product designed for a single life cycle. Thus, there is a strong necessity to move towards life cycle design and encourage quality over quantity in the design of products.

7.4.3 Market Demand

The demand for remanufactured products is affected by customer perceptions of the quality and value of remanufactured products. Ultimately, however, interest in remanufactured products depends heavily on the price of competing new products. Because most customers are neither aware of workmanship that goes into a remanufactured product nor convinced that remanufactured products can perform at as well as their virgin counterparts, they often expect remanufactured products to be sold at a lower price, substantially reducing the margin of remanufacturing and thereby discouraging the growth of remanufacturing industries. A general preference for new products for these reasons reduces the economic viability of remanufacturing despite functional opportunity. This is especially true in electronics and IT product sectors, where rapid innovation and shorter product life cycles create strong competition from new goods that may indeed be technologically superior, with recently-outdated but still advanced "last-generation" new models consistently declining in price [4]. Even though, there is a 'green' and environmentally-conscious customer base that currently supports remanufacturing, reliance on this 'green' customer base alone will not sustain the remanufacturing industry, let alone allow it to grow [3].

7.4.4 Core Supply

Current product supply chain models almost exclusively reflect one-way movement of products from the point of raw material extraction to the point manufacture, and therefrom to customers and ultimately to landfill or incineration. This linear model poses challenges for both OEMs and third party remanufactures in accessing cores at end-of-use (EOU) or EOL because the infrastructure for getting whole *products* back—rather than simply their constituent materials as in post-disposition recycling—is severely underdeveloped [8]. These challenges create uncertainty in the quantity, quality, and timing of core return, making remanufacturing less predictable and thus leading to difficulties in planning remanufacturing operations. For example, when the volume of core return is insufficient, remanufacturers sometimes will have to salvage lower-quality cores or even use new products to meet demand requirements, which can drive up both costs and environmental impacts [9]. Conversely, a surplus of cores creates a need to inventory, adding cost increasing the risk of interim product obsolescence.

Another issue that has been relatively underexplored is the frequency and impacts of damage to cores during transportation processes. In some cases, transport parties perceive cores for remanufacturing as EOL or scrap materials, and their handling practices thus reflect their perception of low value. This can result in increased scrap and fallout rates, as cores were once suitable for remanufacturing become unviable—ultimately damaging the remanufacturer's bottom line. An example of this is the business-to-consumer (B2C) printer cartridge market, where different cores are sometimes collected in a single stream, creating mixed batches that increase the risk of cross-contamination [8].

7.4.5 Skills, Technology, and Data of Remanufacturing

Compared to conventional manufacturing, remanufacturing is considered more labor intensive because, in contrast to highly-automated forward assembly, EOL disassembly and diagnostic processes are overwhelmingly manual. In this context, many decision made in remanufacturing processes rely on engineering experience and intuition, and thus require engineers and technicians whose technical skills extend beyond the norm of controlling forward production systems. However, a lack of skilled talent in a number of remanufacturing sectors is becoming an increasingly prevalent becomes a barrier to the development of remanufacturing. In addition, with the development in materials and manufacturing process, remanufacturers must likewise constantly keep their assessment, repair, and upgrading technologies advanced in order to provide products that can compete with contemporary market offerings. Another concern, particularly relevant with independent remanufacturers, is the availability of product design specifications. In most cases, OEMs are reluctant to share product design and performance data with any party that use it to create competing products, severely limiting the capability of independent remanufacturers to ensure their products achieve the parity that consumer markets expect. A particular challenge in this regard, any regulatory change forcing OEMs to share data may create the risk of encouraging OEMs to offshore production in effort to avoid legislation [8]. Therefore, there is a need to develop a marketplace for the transfer of design specifications and performance data between OEMs and independent remanufacturers to improve information flow on an equitable basis.

7.5 Paving the Way for Uptake of Remanufacturing

The present transaction cost of shifting from current linear business models to a closed-loop model is high, and is therefore beyond the scope of capability of any single group or entity [10]. As a result, unlocking the potential for remanufacturing to transform global economies towards more circular industrial systems requires collaboration across business, government, investors, society, and the research community to address each of these challenges. Fortunately, extensive research efforts have identified a number of key systemic enablers that have the potential to reduce the cost of transformation and increase the broad market acceptance of remanufactured products in this pursuit. This section presents these key enablers as well as several successful case studies to illustrate accelerated adoption of the remanufacturing business model.

7.5.1 Connecting With New Business Models—The Product Service System

Every single night, over 500,000 guests rent accommodation as a service from a single company that offers two million room listings in over 57,000 cities worldwide [12]. Product owners (i.e. homeowners) shared their excess resources—a room—with customers seeking a defined service period, and consequently generated revenue without ever selling or transferring ownership of the physical product. Customers, on the other hand, chose the features and service duration for which they had demand and completed all transactions and communications online. Matching of the product owners and their customers was facilitated by Airbnb, third-party platform and itself a service, that earns revenue by creating successful matches.

Though not a manufacturer in its own right, Airbnb is the most prominent example of a huge new "sharing economy," and the business model upon which it is based is the future of the industrial economy. In this model, the life cycle utility of assets is maximized via collaborative use models, and the life cycle period over which value may be extracted from that utilitiy is similarly maximized through remanufacturing. From sharing bikes to power tools, swapping clothes to leasing unused parking spaces, the sharing economy has gained unprecedented momentum over the past decade, and will soon become the foundation of our everyday life, as graphically shown in Figure 7.2. Indeed, analysis from consultants PricewaterhouseCoopers (PwC) highlights that revenues derived form a sharing economy could rise from $15 billion to over $300 billion by 2025 [13].

Figure 7.2 Illustration of the sharing economy [11].

In the industrial economy, the growth of these sharing models will be driven by the recognition of mutual social, economic, and environmental benefits for both consumers and OEMs. From an economic standpoint, collaborative use models allow asset owners to make profit by simply finding ways to utilize the functional capacity of their products in times when they would otherwise be sitting idle, providing no value. Conversely, such models enable consumers to access to the product/service only as much as they need, reducing the burdens of high capital costs and the loss of value when the need is fulfilled before the functional capacity reached. From environmental point of view, both resources and assets are used in a more efficient and sustainable manner, reducing the unused excess and unnecessary depletion created when customers must purchase much but only need a little. In terms of social impact, connections between people—between asset owners and customers whose roles may constantly swap or merge—is deepened by demanding collaboration communication in the interest of cost, material, and time efficiency.

Currently, there are three types of systems under sharing economy: (1) Redistribution Markets, (2) Collaborative Lifestyles Platforms, and (3) Product Service Systems (PSS). While each are balanced with benefits and challenges, the Product Service System model is the only structure in which consumers are actively involved with producers in a reflection of traditional commercial relationships [14]. By definition, the PSS is "a pre-designed system of product, services, supporting infrastructure, and necessary prearranged network" aimed at fulfilling consumer needs while simultaneously minimizing detrimental economic, social, and environmental impacts [15]. In this, the Product Service System is particularly relevant to the circular economy transition, and is likewise considered as a key enabler of a successful remanufacturing business model.

This model provides an innovative means to cope with the unique complexities that currently inhibit the growth of remanufacturing. Consider, for example, the widely cited challenge of reverse logistics in core return [16]. Traditionally, selling products transfers ownership to the buyer, who is thereafter free to decide the fate of that product after their need is met. In such an ownership system, coordinating product return without well-developed infrastructure requires effort on behalf of the owner that exceeds the effort required in traditional linear disposal pathways. As a result, owners must typically be incentivized to return the product with some benefit at the expense of the remanufacturer. In contrast, the PSS model allows manufacturers to retain ownership, meaning that core return becomes an inherent condition of use rather than a non-mandated and therefore uncertain user decision. Further, the PSS model makes services such as condition monitoring or regular maintenance part of the use agreement rather than an additional service the user may or may not choose to buy. Thus, by ensuring proper maintenance and developing a continuous understanding of condition, the PSS model allows remanufacturers to develop a platform expectation of core quality upon return, reducing the difficulty of diagnostic processes and minimizing the intensity of required remanufacturer interventions. This also increases product longevity and durability, ultimately enabling it to maintain performance over multiple life cycles and thereby extending the period for which value may be derived. Together, these advantages create a degree of stability and predictability in the quantity, quality, and timing of core return, mitigating three of the most significant barriers that currently inhibit the growth of remanufacturing.

Take tires as an example. In traditional ownership-consumption models, consumers use the tires for an undetermined amount of miles until they are effectively unsalvageable, and thereafter purchase replacements at significant economic and environmental cost. In a theoretical PSS, consumers could purchase a defined mileage of use through leasing at fixed rate and subsequently return used tires to nearby workshops for re-treading. If tires fail before the defined use mileage, consumers could have them repaired directly in order to fulfil the remaining expected life if the tire. Within this system, previously under-used but highly valued assets can be greater utilized rather than replaced, driving down the associated operating cost per unit of use. Service providers could also have a better control of the material, component and product, which lowers maintenance cost and increases product longevity, enhancing the margin for revalorizing products after each usage cycle. Importantly, such a system would be viable for deployment by both OEMs and independent remanufacturers, providing equitable market opportunity.

PSS comprises both tangibles (products) and intangibles (services) to fulfil specific customer needs. Based on the weight combination of tangibles and intangibles, three classes of PSS have been identified (Figure 7.3).

1. **Product-Oriented**: the product is sold to the customer together with additional services contract (e.g. maintenance) to ensure the functionality of the product. Value proposition is still primarily

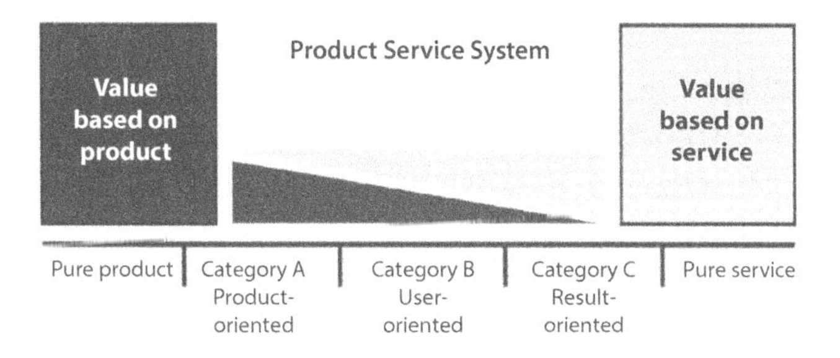

Figure 7.3 The product-service concept [17].

based on selling (i.e. transferring ownership) of the product.

2. **User-Oriented**: the product is made available to consumer through various contractual agreements, such as sharing, pooling, and leasing. As ownership of the product remains with service provider, the value proposition will rely on both product and service.
3. **Result-Oriented**: selling the product is replaced completely by selling service; both supplier and consumer agree on the service delivered. This is considered the most widely-applicable definition with greatest potential for a PSS model.

Several big market players have identified the potential economic benefits of PSS and incorporated the PSS model to increase their competitive advantages. For example, Fuji Xerox sells the service of copying paper based on the number of copies instead of selling entire copy machines. Similarly, Volvo Aero engine providers charge for maintenance, repair, and overhaul services per engine flying hour rather than selling complete replacement aircraft engines. Likewise, Philips lighting provides a defined level of illumination in a building, charging based on a per-lux (unit of illumination) model rather than simple selling light bulbs.

For OEMs to offer PSS model, they must have intrinsic motivations to extend product lifetime as long as possible. In the context of remanufacturing, this typically stems from an ability to continue deriving value from the product's function as long as it continues to perform, which encourages recovery and remanufacturing as a means to minimize the material, energy, labor, and opportunity costs of new product manufacturing. In addition, because ownership remains with the OEM, companies are better enabled to keep track of product performance during use, predict core supply, and thus plan remanufacturing operations efficiently. Recognizing the potential economic opportunity in this structure will likely encourage decision makers to design products for longer life cycles and for remanufacturing, making it easier and more economically attractive compared to constant virgin production in pursuit of unit sales. From the consumer perspective, the utility of ownership itself will decrease, as the value for which consumers pay is inherently in the service rather than the physical components of the product. As a result, the "need" to own a new product will decrease, and acceptance of remanufactured goods will increase based on their provision of valuable function, subsequently supporting the growth of remanufacturing business models. In this system where service *is* the product and asset health serves the interest of the OEM as well, consumers could be guaranteed access to service either as a component of use or at competitive price, without worrying about the maintenance and/or repair cost that might be incurred for user-owned products which, in traditional systems, OEMs have little interest in preserving [18]. Moreover, research suggests that economic advantages in leasing are available to both OEMs and consumers based purely on the avoided incurrence of high capital costs for either producing or purchasing a virgin (new) product.

Interestingly, firms show more interest in remanufacturing as a necessary component of a PSS model, than in offering remanufacturing services alone as a complement to sales-based businesses. Indeed, a survey of 625 senior manufacturing decision-makers in 13 countries found that 86-percent of them view the transition from product-oriented to service-oriented revenue models as a core strategy for development, reflecting immense opportunity for the uptake of remanufacturing [19]. For now, more than 50-percent of respondent manufacturers include service as a part of their product-based business model. This population may serve as a sound potential target for initially promoting PSS models as a means by which to achieve natural dissemination of remanufacturing as a competitive business practice.

7.5.2 Setting Up Global Reverse Supply Chain

Driven by the growing emphasis on environmental issues, resource scarcity, and ongoing economic volatility, many OEMs are increasingly interested in extending their involvement in the product life cycle past the point of delivery. To achieve this requires advanced networks for tracking and handling products at the EOU or EOL back from the user to the manufacturer. Such networks, known as reverse or closed-loop supply chains, necessarily depend on reverse logistics activities, "a specialized segment of logistics focusing on the movement and management of products and resources after sale and delivery to the customer" [20]. Reverse logistics systems are not presently well-established in most industries, but are continuously undergoing development. Based on the geography and leakage of material post-use, three archetypes of circular or partially-circular supply chain organization have been identified: (1) global/regional closed loop, (2) partially open local/regional loop, and (3) open cascade, as summarised in Figure 7.4 [10].

Ideally, the closed-loop supply chains will stay within regional or local areas in order to minimize transportation costs and international trade logistics. While this is desired in the classical sense, increasing raw material prices and the efficiency of transportation and logistics systems have created great opportunity for a large-scale arbitrage economy that extends product and material recycling to global scale. Ricoh, for example, captures this opportunity by shipping low-value recycled plastic

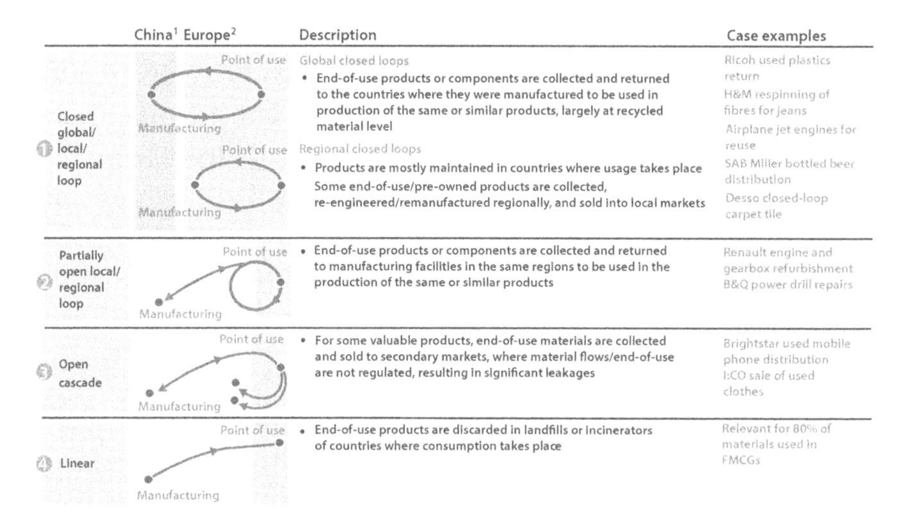

Figure 7.4 Archetypes of reverse supply chain and closed loops [10].

from recovery sites in Europe to Asia, where it holds greater value as a feedstock for manufacturing their new products. While this global scale does not fit the classical definition, it is still effectively closed-loop; even accounting for transportation costs, Ricoh is still able to achieve 30% savings in material cost over purchasing virgin plastic feedstocks [21]. In this sense, it is expected that as trade is increasingly globalized and access to resource becomes more constrained, reverse logistics and closed-loop supply chains may benefit from economies of scale in much the same way as their linear predecessors. One reflection of this is the launch of the world's largest container ship, Triple-E, in 2013, which surpassed the capacity of the previous record holder by 16% while reducing CO_2 emissions by about 35-percent per container moved [21]. As the economic and environmental costs if transportation decrease, the representative portion of transportation in total cost will similarly decline, better enabling companies to explore globalized reverse logistics as a triple-bottom-line serving measure.

It must be recognized, however, that reverse logistics is not necessarily a symmetric picture of forward supply chain; rather, it demands unique operations of its own [22]. Remanufacturing, of course, starts with collecting cores from users and transporting them back to the remanufacturing facility. However, variations in use contexts, product life durations, owner retention, and change in technology often creates uncertainty in the quantity, quality, and timing of this return [21–23]. An absence of developed return infrastructure compounds this uncertainty, as users are typically most inclined to choose the easiest route of disposition. Handling this uncertainty is critical to creating profitable remanufacturing business models, as it will directly change the inventory and operation planning obstacles that inhibit growth.

To reduce this uncertainty, remanufacturers can actively manage the process of core collection and strategize the acquisition process. Beside leasing models in which OEMs retain ownership, other strategies include the deposit-based approach, in which customers pay a deposit upon purchase that incentivizes them to return the product at the EOU to reclaim that fee; an exchange-based model, in which customers are obligated to return similar used products as a prerequisite to purchasing a replacement remanufactured product; the credit-based approach, in which customers receive credit for returning cores; and the buy-back approach, where remanufacturers simply purchase cores from brokers or scrap yards [9]. Caterpillar Inc., for example, has adopted a credit-refund policy with their customers to ensure the quantity and quality of the returned cores. When customers return the used components, Caterpillar will inspect their condition and determine variable credit amounts based on quality [24]. Companies seeking to build successful reverse supply chains for remanufacturing must properly choose and/or design the types of core acquisition relationships that best match their product profile.

Besides core acquisition model, the next challenge is to design and plan the actual transport networks to get components from the point of use back to the remanufacturing facility. Distribution shipments in forward logistics usually involve high volume and relative homogeneity in product type, transporting goods from manufacturers to a few local clients or dealers. In a reverse supply chain, however, cores available for collection at any given time are markedly lower in volume and unavoidably diverse in type, both of which may vary considerably. This creates significant challenges in achieving cost-effectiveness [25]. In this regard, sophisticated reverse network management capabilities comprised of both hardware (e.g. the collection points/channels) and software (e.g. condition monitoring of used component, inventory management, sorting standards, network planning, and core forecasting capability) must be established. Further, companies must also design product packaging systems carefully to ensure the quality of core return—an often overlooked consideration. If cores are not packaged well, they are more susceptible to damage, which may reduce their recoverable value.

Building effective reverse logistics channel is essential for the success of remanufacturing business. Observation of current practices suggests that while recycling of raw materials can be conducted at a global scale, core collection networks for remanufacturing are likely to function best at local or regional scales due to the notably variable volume and type factors, as well as the physical complications of transporting complete products. However, the economy-of-scale effect emerging in global closed-loop supply chains, as described above, haw to some extent introduced opportunity to expand the viability of remanufacturing core collection networks to broader scales. Companies must carefully evaluate the particularities of their industries to create new opportunities and prepare for the global uptake of remanufacturing.

7.5.3 Innovative and Enabling Technology from Industry 4.0

The term industry 4.0 refers to the fourth stage of industrialization which aims for realizing a high level of automation in manufacturing through electronics and Information Technology. As this becomes a reality, the boundaries between the real world and the virtual world are increasingly blurred, leading to the development of integrated cyber-physical production systems (CPPSs). In other words, mechanical and electronic elements of production systems—and even products themselves—are linked by IT and communicate with each other [26]. In the Industry 4.0 environment, machines, products, ICT systems, stakeholders, and decision-makers across the entire value chain and the complete product life cycle are linked with each other, through smart networks (Figure 7.5). The development towards Industry 4.0 has not only presented a substantial opportunity for the traditional manufacturing industry, but also brought along opportunities for closing material loops and catalysing the development of remanufacturing. New enabling technologies can be best classified into three key areas:

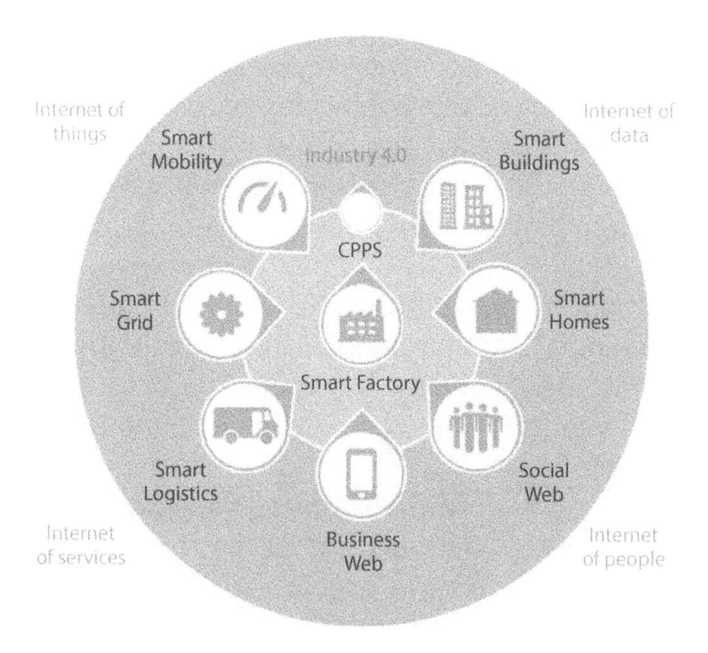

Figure 7.5 The Industry 4.0 environment [26].

Smart data for greater connections

In the year 2010 alone, there were already more "things" connected to the internet than people on earth—over 12.5 billion devices. This number is expected to grow to 50 billion by 2020 [27]. PCs, Laptops, smart phones, machines, ATM machines, watches, and in the near future, even furniture, automotive parts, tires, clothes, air conditioners—indeed, nearly everything—is likely to be connected via sensors and networks. Data captured in real-time can be stored, telling the digital "life-story" of everything from whole products down to their constituent components and materials; enabling previously inconceivable levels of tracking efficiency. The astonishing rise in the data volume, computational power, and connectivity will facilitate the reintegration of EOL products and materials into central economic systems. Making use of captured data through computational analysis, service providers may monitor the condition of products in real-time and optimize the return of the products for predictive maintenance or remanufacturing before any catastrophic failure happens [28]. For example, RFID technology may be used to track the whereabouts and use condition of an HDOR engine component offering insight and supporting optimized EOL decision making for reuse, remanufacturing or recycling. The greater data availability and computational power will also accelerate the interest of stakeholders and encourage new and disruptive business models, such as PSS.

Additive manufacturing for pars repair and replacement

Additive manufacturing is another disruptive technology that makes Industry 4.0 a reality. It is the process of joining materials to make objects from computerized 3D model data, usually layer upon layer, as opposed to subtractive manufacturing methodologies [29]. Additive manufacturing provides an automated and precisely-controlled approach to rework worn parts. For example, when aerospace engine blades get damaged, traditional remanufacturing methods necessitate disassembling and restoring the entire blade assembly. In contrast, a Laser Metal

Deposition technique could be used to melt metal powder and deposit fusion-bonded metal directly onto the blades, restoring the original geometry and material properties without excessive disassembly, subtractive surface repair, or replacement. Further, this can all be done using a Computer Numerical Control (CNC) robot on a gantry system, minimizing human workplace risks and labor costs. This level of automation and precision is particularly important when repairing especially fine parts with minute tolerances of complex geometries, such as blade tips of 0.2mm thickness where manual welding is virtually impossible. In addition to restoration, additive manufacturing also holds great opportunity for reducing inventory costs by enabling on-demand creation of small volume, difficult-to-find, or expensive parts.

Advanced remanufacturing technologies

Evolving technologies and processes such as laser scanning, non-destructive testing, laser cleaning, and adaptive machining, have continuously driven down the cost of remanufacturing operations and improved achievable part quality. Consider, for example, the advent of adaptive machining. Adaptive machining starts with scanning a part to account assess its unique geometry and build a tailored machining program to fit it specifically. Using standard baseline building blocks for machining processes, multiple parts with different geometries can all be machined to the same specifications despite their variations without complex reprogramming or retooling measures. In other words, adaptive machining provides the capability to machine multiple complex but marginally different components with a single process and in a completely automated manner. This can allow remanufacturers to cope with the variation in product use contexts (and thus core quality) that often create uncertainty in the remanufacturing process. Thanks to the increasing computing power to process large amounts of data, the complex computation for adaptive machining can be performed in real time and the CNC machine can handle the complex machining data even at high feed rate. Advanced technologies such as this are critical in expanding the portfolio of capabilities with which remanufacturers are equipped, and thus in supporting innovation central to competition and growth.

7.5.4 Design for Remanufacturing

The decision to include remanufacturing as a part of product life cycle considerations should be made early in the business-case stages of product development that inform design choices, as about 80-percent of a product's costs are determined by its design [30]. Thus, if proper consideration is given to design products deliberately to enable and encourage remanufacturing, many of the barriers that presently inhibit its growth can be mitigated [2, 31–34]. The underlying philosophy in this pursuit, however, depends on the understanding that product designers are not independent actors. That is to say, designers create product concepts that meet a defined set of parameters for cost and performance that are themselves based on business goals, market need, and manufacturing infrastructure. Thus, designing for remanufacturing must be a systemic decision in which remanufacturability is highlighted as a necessary product parameter, rather than an optional, nonconsequential design choice. In this light, every stage of product development must be informed with a base of remanufacturing knowledge so that features of remanufacturability may be integrated fundamentally into the product concept, and thereby ease the EOL remanufacturing process. An important consideration in this regard is that contracted and independent remanufacturers—perhaps the largest portions of the industry—have no control over these upstream stages, and are only able to remanufacture products that already exist. It is essential to recognize, therefore, that the power to inhibit or unlock remanufacturing ultimately lies with the OEMs responsible for product design and development, regardless of whether or how much they engage in remanufacturing themselves.

From a holistic level, design for remanufacturing comprises two levels of design activities: designing the business model and designing the product.

In terms of designing the business model, the opportunity for remanufacturing must be carefully evaluated to determine whether remanufacturing should be included in the product support portfolio. Factors that must be evaluated and analyzed usually include:

1. **Laws and regulations**, which examine the impacts of and restrictions created by environmental legislation with respect to remanufacturing
2. **Market demand/acceptance**, which evaluates the extent to which the product is expected to occupy a competitive position in the market after being remanufactured
3. **Return potential**, which predicts the likelihood that a product will be successfully returned to the remanufacturing site at the end of its useful life
4. **Remanufacturing capability**, which analyzes the technology and information required to remanufacture a product
5. **Economic incentive**, which predicts value that may be recovered by remanufacturing
6. **Organizational support and consideration**, which aim to understand whether remanufacturing decisions will be understood and supported by upper management

In terms of designing the product for remanufacturing, product developers must aim first and foremost to increase the value that can be recovered from remanufacturing. Usually, this is accomplished through "Design for X" strategies, where X may be any post-use process—such as disassembly, core recovery, cleaning, or upgrading—that can be made easier with proper choices upstream in the product life cycle. The prioritization of these design activities may vary according to the requirements and capabilities of the remanufacturers themselves. Usually, design guidelines—simplified, directional, and qualitative frameworks for design decision-making—are the most straightforward and commonly used approach to Design for Remanufacturing [33, 34]. Examples include broad objectives such as reducing the number of joins and fasteners, using durable materials, avoiding using permanent joints, etc. Following such guidelines in any particular pursuit (be it disassembly, durability, or anything else) inherently steers designs toward greater remanufacturability.

Xerox Corporation is perhaps one of the most well-known practitioners of design for remanufacturing. For many years, the company has developed assessment tools to carefully evaluate the feasibility of remanufacturing their industrial printers, and strategized the business model for remanufacturing through initiatives such as leasing machines rather than selling them. Once the economic and technical opportunity were clear, Xerox began to deliberately design their printers with modularity and with fewer parts, aiming to ease the disassembly process and enable component replacement rather than complete asset disposition. The designers also used a common platform approach, especially for the main engine of the machine and peripheral equipment, in which the same central technologies are used across different product models, minimizing amount of unique parts across their product range and reducing the vulnerability to differential obsolescence. Research focusing on Xerox as a case study suggests that by incorporating remanufacturing as a central principle of product *and* business model design, the company effectively unlocked seven additional revenue streams from products remanufactured to serve second through seventh useful life cycles [35].

7.6 Conclusion

In an era where demand for finite resources will only continue to rise, the way people produce and consume goods must change;

in short, the global community must transition from linear systems of extraction, production, consumption, and disposal to a more circular economy. As an effective closed-loop measure in circular economy, remanufacturing presents immense potential for economic, environmental, and social benefits. However, global uptake of remanufacturing is still hampered by a number of challenges in standards and legislation, design awareness, market demand, uncertainty in core collection, labor availability, technical data flow, and remanufacturing process technology. To address these barriers is simply beyond any single entity's scope of capability; rather, large scale adoption of remanufacturing in pursuit of a circular economy requires collaboration amongst multiple players across business, research, government, and investor communities. A number of enablers that present the opportunities to reduce the cost of transformation towards a circular economy—improving the technical feasibility, economic profitability, social benefit, and environmental impacts of remanufacturing—are now clearly identified and must be pursued. Fortunately, the need to action and the capability to take action have never been better aligned. By actively working to increase market adoption of remanufacturing and remanufactured products, we may effectively pave the way for an increase in global uptake of new remanufacturing strategies that can decouple the increasing need for socioeconomic development from its conventional reliance on increased resource use.

References

1. Guidat, T., Barquet, A.P., Widera, H., Rozenfeld, H., Seliger, G., Guidelines for the definition of innovative industrial product-service systems (PSS) business models for remanufacturing. *Procedia CIRP*, 16, 193–198, 2014.
2. APSRG & APMG, Triple Win - The Social, Economic and Environmental case for Remanufacturing, http://www.policyconnect.org.uk/apsrg/sites/site_apsrg/files/triple_win_-the_

social_economic_and_environmental_case_for_remanufacturing.pdf, 2014.
3. Sundin, E., Product and process design for successful remanufacturing. Linköping Studies in Science and Technology Dissertation, No. 906, 2004.
4. US International Trade Commission (USITC), Remanufactured goods: an overview of the U.S. and global industries, markets, and trade, http://www.usitc.gov/publications/332/pub4356.pdf, 2012.
5. Construction Shows. *4th China International Remanufacturing Summit*, 2014.
6. Cnenergy, overview of remanufacturing market in China in year 2016, http://www.cnenergy.org/hb/201606/t20160614_318091.html, 2016.
7. Matsumoto, M., Yang, S., Martinsen, K., Kainuma, Y., Trends and research challenges in remanufacturing. *Int. J. Precis. Eng. Manufact.-Green Technol.*, 3, 1, 129–142, 2016.
8. ERN, Remanufacturing Market Study, https://www.remanufacturing.eu/wp-content/uploads/2016/01/study.pdf, 2015.
9. Sundin, E. and Dunbäck, O., Reverse logistics challenges in remanufacturing of automotive mechatronic devices. *J. Remanufacturing*, 3, 1, 1–8, 2013.
10. Waughray, D., Preface – World Economic Forum, http://reports.weforum.org/toward-the-circular-economy-accelerating-the-scale-up-across-global-supply-chains/preface-world-economic-forum/, 2014.
11. Bacon, D., The sharing economy, http://www.illustrationweb.com/sg/artists/DerekBacon/all/540, 2013.
12. DMR, By the number: 47 Amazing Airbnb statistics, http://expandedramblings.com/index.php/airbnb-statistics/, 2016.
13. PwC Blog, Five key sharing economy sectors could generate £9 billion of UK revenues by 2025, http://pwc.blogs.com/press_room/2014/08/five-key-sharing-economy-sectors-could-generate-9-billion-of-uk-revenues-by-2025.html, 2014.
14. Botsman, R. and Rogers, R., *Beyond Zipcar: Collaborative Consumption*, Harvard Business Review, https://hbr.org/2010/10/beyond-zipcar-collaborative-consumption/, Boston, USA, 2010.

15. Guidat, T., Barquet, A.P., Widera, H., Rozenfeld, H., Seliger, G., Guidelines for the definition of innovative industrial product-service systems (PSS) business models for remanufacturing. *Procedia CIRP*, 16, 193–198, 2014.
16. Mont, O., Clarifying the concept of product-service system. *J. Clean. Prod.*, 10, 3, 237–245, 2002.
17. Sundin, E., Östlin, J., Rönnbäck, A.Ö., Lindahl, M., Sandström, G.Ö., *Remanufacturing of products used in product service system offerings, Manufacturing Systems and Technologies for the New Frontier*, pp. 537–542, Springer, London, 2008.
18. Mont, O., Dalhammar, C., Jacobsson, N., A new business model for baby prams based on leasing and product remanufacturing. *J. Clean. Prod.*, 14, 17, 1509–1518, 2006.
19. Hilson, G., Manufacturers must move to service-oriented business models, leverage IoT: Cisco, http://www.itworldcanada.com/article/manufacturers-must-move-to-service-oriented-business-models-leverage-iot-cisco/378543#ixzz4AQvqwLkx, 2015.
20. Andrew, K.R., Meeting the Reverse Logistics Challenge. *Supply & Demand Chain Exec.*, 6, 44–47, 2005.
21. World Economic Forum, Towards the circular economy: Accelerating the scale-up across global supply chains, http://reports.weforum.org/toward-the-circular-economy-accelerating-the-scale-up-across-global-supply-chains/, 2014.
22. Fleeischmann, M., Reverse logistics network structure and design, repub.eur.nl/pub/113/erimrs20010919163815.pdf, 2001.
23. Guide, V.D.R., Production planning and control for remanufacturing: Industry practice and research needs. *J. Oper. Manag.*, 18, 4, 467–483, 2000.
24. Fleischmann, M., Krikke, H.R., Dekker, R., Flapper, S.D.P., A characterisation of logistics networks for product recovery. *Omega*, 28, 6, 653–666, 2000.
25. Östlin, J., Sundin, E., Björkman, M., Product life-cycle implications for remanufacturing strategies. *J. Clean. Prod.*, 17, 11, 999–1009, 2009.
26. Caterpillar Inc, Core acceptance criteria, http://www.caterpillar.com/en/company/sustainability/remanufacturing/advantages.html, 2016.

27. Barquet, A.P., Rozenfeld, H., Forcellini, F.A., An integrated approach to remanufacturing: Model of a remanufacturing system. *J. Remanufacturing*, 3, 1, 1–12, 2013.
28. Deloitte, Industry 4.0 Challenges and solutions for the digital transformation and use of exponential technologies, http://www2.deloitte.com/content/dam/Deloitte/ch/Documents/manufacturing/ch-en-manufacturing-industry-4-0-24102014.pdf, 2015.
29. Evans, D., The Internet of Things, How the Next Evolution of the Internet, Is Changing Everything, http://www.cisco.com/c/dam/en_us/about/ac79/docs/innov/IoT_IBSG_0411FINAL.pdf, 2001.
30. Stock, T. and Seliger, G., Opportunities of sustainable manufacturing in Industry 4.0. *Procedia CIRP*, 40, 536–541, Boston, USA, 2016.
31. David, M. and Anderson, P.E., *Design for Manufacturability: How to Use Concurrent Engineering to Rapidly Develop Low-Cost, High-Quality Products for Lean Production*, CRC Press, Florida, USA, 2014.
32. Nasr, N. and Thurston, M., Remanufacturing: A key enabler to sustainable product systems. *Proc. of 13th CIRP International Conference on Life-Cycle Engineering*, pp. 15–18, 2006.
33. Amezquita, T., Hammond, R., Salazar, M., Bras, B., Characterizing the remanufacturability of engineering systems. *Proc. of ASME Advances in Design Automation Conference*, pp. 271–278, 1995.
34. Ijomah, W., McMahon, C., Hammond, G., Newman, S., Development of robust design-for-remanufacturing guidelines to further the aims of sustainable development. *Int. J. Prod. Res.*, 45, 18–19, 4513–4536, 2007.
35. Charter, M. and Gray, C., Remanufacturing and product design. *Int. J. Prod. Dev.*, 6, 3–4, 375–392, 2008.

Index

CPSIA information can be obtained
at www.ICGtesting.com
Printed in the USA
BVHW042038131019
560894BV00008BA/65/P